The Business of New Process Diffusion

The Business of New Process Diffusion explores entrepreneurship, innovation and process diffusion through the example of the development of float glass. The significance of the glass industry as a vehicle for studying innovation activities has been recognised for some time. By using it as an example to draw out the key themes of innovation and diffusion theory, this book uses its specific industrial history to form an illuminating case study.

Little has been written in terms of the management of the early float glass start-ups, resulting in a gap in the literature. This book seeks to remedy this by recounting developments through the lens of one of the leading glass technologists involved in the process at the time, using historical and archival material, and artefacts from the period. It illustrates the business origins of the process and its invention, progressing to innovation, competition in the market, first successful production, licensing and patents, and the management of the start-ups leading to market leadership: all significant to the study of technology, entrepreneurship and innovation.

This short-form volume provides a concise but rich resource for researchers and students of the theory and practice of innovation, new process diffusion and start-up management.

Brychan Celfyn Thomas is a Visiting Professor in Innovation Policy at the University of South Wales, a Doctoral Supervisor in Entrepreneurship and Innovation at the University of Gloucestershire, and an Examiner for the International Baccalaureate Business and Management Diploma. Before retiring in October 2012, he was Reader in Innovation Policy at the University of Glamorgan Business School.

Alun Merlyn Thomas was educated at King Williams College in the Isle of Man and holds a BA (Hons) in Economics from Birmingham City University and is an Associate Member of the Chartered Institute of Secretaries & Administrators (ACIS). He was for some years a member of the Institute's Branch Council for South Wales. By way of occupation he has been a company accountant for a number of engineering organisations.

Routledge Focus on Business and Management

The fields of business and management have grown exponentially as areas of research and education. This growth presents challenges for readers trying to keep up with the latest important insights. Routledge Focus on Business and Management presents small books on big topics and how they intersect with the world of business research.

Individually, each title in the series provides coverage of a key academic topic, whilst collectively, the series forms a comprehensive collection across the business disciplines.

The Neuroscience of Rhetoric in Management
Compassionate Executive Communication
Dirk Remley

Heidegger and Entrepreneurship
A Phenomenological Approach
Håvard Åsvoll

The Politics of Organizational Change
Robert Price

Globalization and Entrepreneurship in Small Countries
Mirjana Radović-Marković and Rajko Tomaš

The Business of New Process Diffusion
Management of the Early Float Glass Start-ups
Brychan Celfyn Thomas and Alun Merlyn Thomas

For more information about this series, please visit: www.routledge.com/
Routledge-Focus-on-Business-and-Management/book-series/FBM

The Business of New Process Diffusion

Management of the Early Float Glass Start-ups

Brychan Thomas

**Brychan Celfyn Thomas and
Alun Merlyn Thomas**

 Routledge
Taylor & Francis Group

LONDON AND NEW YORK

First published 2019
by Routledge
2 Park Square, Milton Park, Abingdon, Oxon OX14 4RN

and by Routledge
52 Vanderbilt Avenue, New York, NY 10017

Routledge is an imprint of the Taylor & Francis Group, an informa business

British Library Cataloguing-in-Publication Data
A catalogue record for this book is available from the British Library

Library of Congress Cataloging-in-Publication Data
A catalog record for this book has been requested

ISBN: 978-1-138-58719-9 (hbk)
ISBN: 978-0-429-50410-5 (ebk)

Typeset in Times New Roman
by Apex CoVantage, LLC

MIX
Paper from
responsible sources
FSC
www.fsc.org   FSC™ C013985

Printed in the United Kingdom
by Henry Ling Limited

In memory of James Edward Celfyn Thomas

MA Cantab, CEng, FIMechE (1925–2005)

Float Glass Manufacturing Manager at Pilkington Brothers' Cowley Hill Works, where he started up the world's first float glass plant in 1957.

Source: Triplex Magazine, Vol. 20, No. 12, December 1966, p. 227

Contents

Figures

Tables

Abbreviations

DCF	Discounted cash flow
HTFs	High-technology firms
ICT	Information and communication technology
IP	Intellectual property
LOF	Libbey Owens Ford
OECD	Organisation for Economic Co-operation and Development
PPG	Pittsburgh Plate Glass
R&D	Research and development
SME	Small and medium-sized enterprise
StG	St Gobain Compagnie de St Gobain
UK	United Kingdom
USA	United States of America

Foreword

The float glass process is a classic example of a major, industry-transforming process innovation. This book plots the history of how the innovation came about and sets it carefully in its historical and conceptual context. It is a rich case study told with the aid of new material from the archives of somebody actually involved at the time and illustrates many aspects of the theory and practice of innovation management.

The float glass process is so strongly associated with the name of Pilkington that it comes as a surprise that the core idea had been invented, and patented, fully 50 years before Pilkington reinvented it and turned it into a practical proposition. It's a perfect example of how innovation is so much more than mere invention. How much more is shown simply by the fact that it cost the company £7 million to turn the invention into something useful – that is, into an innovation. Some scholars debate whether small companies are intrinsically more innovative than large ones. There is no clear answer, and anyway it is a somewhat sterile debate. But certainly, only a substantial company with significant, well-developed expertise could make such an investment. It would be stretching a point to view this case as an early example of open innovation, but it is certainly an example of successfully adopting an idea from elsewhere and putting it to good use. One is reminded of the (alleged) NIHBWGWUTIA award offered by a company to encourage people to look outside the company as well as inside to find useful ideas. It stands for "Not Invented Here But We're Glad We Used The Idea Anyway".

Remarkably for a company that was already 75 years old, Pilkington had a record of innovations before the float glass project began and clearly must have had a culture that was quite open to change. Nevertheless, all innovations generate resistance, and not just because people are apathetic or want a quiet life. Often they may have quite rational fears that if they are closely involved with the innovation and it fails they will share the blame; or if it is a success, perhaps their own expertise will no longer be as useful as it was and they face being sidelined or worse. The plate glass process reduced manufacturing costs by at least a quarter, making many roles redundant and changing the whole manufacturing organisation from a batch process to a flowline. A theme well illustrated here is the important role of the innovation champion in promoting and nurturing the new process. As the development costs ramped up, Alastair Pilkington's resolve and influence were essential to maintain support throughout the company when many must have felt threatened.

Business model innovation has been much discussed in recent years. Here again the case is pertinent because although the initial aim was to improve the core process on which the company depended, a major part of its commercial success came from licensing the process to potential competitors. This new business model not only provided very substantial additional income but tied competitors to the process and removed their incentive to undertake parallel developments that might have posed a long-term threat. This was done by providing help and support to the start-up activities and setting up a regime in which both parties shared improvements as they were made. This, indeed, makes the float glass an early example of open innovation. What a rich and illustrative case study this is!

R. F. Mitchell
Institute for Manufacturing, University of Cambridge

Preface

The idea to write about the management of the early float glass start-ups evolved more than 12 years ago. At this time the authors discussed the possibility to do this, and when the opportunity arose about a year ago they were more than pleased to commence this book project. In fact, it is opportune and poignant to do so since the current time marks 60 years since the float glass process was commercialised in the late 1950s. Reflecting on the history and development of the process is fascinating especially when recounting events, but also emotional, remembering times when people were striving to create a technological breakthrough. The route taken in this book has been to move along a continuum from innovation management (Cortimiglia et al., 2015; Goffin and Mitchell, 2016), through diffusion (Thomas, 1999) to start-up.

At times the development of the process required working through the night on the tanks, and also with the start-ups, involved much dedication. In these sorts of situations it can be recounted, as an academic once said – you should always keep a notepad and pen next to your bed in case you have an important idea in the middle of the night so that you can write it down and remember it the next day. This is the sort of approach that was taken with one of the float glass developers, who even had a blackboard and chalk in the bedroom!

This book is based on the work of James Edward Celfyn Thomas, who was a key member of Sir Alastair Pilkington's team, and was one of the great glass manufacturing managers and glass technologists from this period. Celfyn Thomas was "float glass manufacturing manager at Pilkington Brothers' Cowley Hill Works, where he started up the world's first float glass plant in 1957" and was "responsible for establishing float glass plants in Maryland, for the Pittsburgh Plate Glass Company, and in California, France, Belgium and Japan. He was senior production manager with a team which designed the St. Gobain float plants at Pisa, Italy, and Porz, near Cologne; also a Ford plant in Tennessee" (Triplex, 1966). When Celfyn sadly passed away in March 2005, David Pilkington wrote to the authors,

> It was one thing to have the idea, it was quite another to make it work. But to make it work on a pilot plant was quite another thing to turn it into a commercial production. This is where your father came in, What a man!

In a further extract from the letter David Pilkington said "Alastair would visit at least twice a day and decide what to do next, but your father had to stay there all the time sometimes day and night and, of course, weekends making it work" (Pilkington, 2005).

For a process innovation to be successful it has to be operational, and this is especially the case with a glass production process as evidenced with the float glass process and start-ups. Without the production of high-quality glass, there would not have been an innovation. This is where those who worked on the float glass process came to the fore. It was an era of self-less dedication by these glass experts, involving inspiration, motivation and insight (Buheji and Thomas, 2016).

This study recounts important developments concerning the float glass process through the "lens" of Celfyn Thomas, as one of the leading glass technologists involved in the process at the time, using historical and archival material, and artefacts from the period. By doing this rather than just providing theorised findings developing established theories based on secondary information, with regard to new process development and start-ups, this book provides a rich source of information using archival material woven into the findings of previous studies to provide greater understanding of what actually happened to make the process a success. This shows that it was not only important for the process to be invented and developed as an innovation but also for it to be made to work effectively to produce high-quality glass and to be diffused through the early float glass start-ups.

Brychan Celfyn Thomas and Alun Merlyn Thomas
Cardiff, December 2018

References

Buheji, M. and Thomas, B. (2016) *Handbook of Inspiration Economy*, Copenhagen: Ventus.

Cortimiglia, M. N., Delcourt, C.I.M., De Oliveira, D. T., Correa, C. H. and Danilevicz, A. de M. F. (2015) A systematic literature review on firm-level innovation management systems, *International Association for Management of Technology, IAMOT 2015 Conference Proceedings*, Cape Town, June 8–11.

Goffin, K. and Mitchell, R. (2016) *Innovation Management: Effective Strategy and Implementation*, 3rd edn, London: Palgrave Macmillan.

Pilkington, D. (2005) *Letter to Alun Thomas* (see Appendix 1).

Thomas, B. (1999) A model of the diffusion of technology into SMEs, *Proceedings of the 44th International Council for Small Business (ICSB) World Conference: Innovation and Economic Development*, Naples, June 20–23.

Triplex (1966) Mr. J.E.C. Thomas, *Triplex Magazine*, December 12, p. 20.

Biographies of the authors

Brychan Celfyn Thomas is a Visiting Professor in Innovation Policy at the University of South Wales, a Doctoral Supervisor in Entrepreneurship and Innovation at the University of Gloucestershire and an Examiner for the International Baccalaureate Business and Management Diploma. He was educated at King Williams College in the Isle of Man and has a Science degree and an MSc in the Social Aspects of Science and Technology from Aston University and a PhD in Science and Technology Policy, CNAA, London. During the academic year 2008–2009 he was on secondment as a Fellow of the Advanced Institute of Management at the Centre for Technology Management, University of Cambridge. Before retiring in October 2012, he was Reader in Innovation Policy at the University of Glamorgan Business School.

He has over 400 publications in the areas of science communication, entrepreneurship, innovation and small business – including 141 refereed journal articles and 140 refereed conference papers. His books include *Triple Entrepreneurial Connection: Colleges, Government and Industry* (2000); *E-Commerce Adoption and Small Business in the Global Marketplace: Tools for Optimization* (co-editor, 2010); *Innovation and Small Business*, Volumes 1 and 2 (co-editor 2011); *Technology-Based Entrepreneurship* (2013); *Academic Working Lives: Experience, Practice and Change* (co-editor, hardback 2014; paperback 2015); *Financial Entrepreneurship for Economic Growth in Emerging Nations* (co-editor, 2018); *Exploring Consensual Leadership in Higher Education through Co-operation, Collaboration and Partnership* (co-editor, 2018) and Innovation and Social Capital in Organizational Ecosystems (co-editor, 2019). He is currently co-leader with Dr Lyndon Murphy of the Leadership, Innovation, Management and Entrepreneurship (LIME) Research Group based in South Wales and is a Fellow of the Chartered College of Teaching and a Fellow of the Higher Education Academy. He has successfully supervised 11 PhDs and two DBAs in the areas of innovation and small and medium-sized enterprises.

Alun Merlyn Thomas was educated at King Williams College in the Isle of Man and holds a BA(Hons) in Economics from Birmingham City University and is an Associate Member of the Chartered Institute of Secretaries & Administrators (ACIS). He was for some years a member of the Institute's Branch Council for South Wales. By way of occupation he has been a company accountant for a number of engineering organisations. He has a special interest in the development and operating of SMEs from a financial perspective and historical opera recordings. His publications include 'Who audits auditors' – letter to the Times – March 1978; 'Bass voices' – letter to Gramophone – April 1979 Vol 56; 'Shorts' – letter to Gramophone – October 1987 Vol 65, and 'King Arthur & The Secret Code' – C & B Thomas – Olfran Books – October 1992 – researcher.

Acknowledgements

We would like to thank our families, especially Anne Thomas, Liz Thomas and Joshua Thomas, for their support during the process of compiling the source material and the writing of this book. Also, we thank our friends, colleagues and other interested parties who have encouraged us to recount the story of the early float glass start-ups and the role played by James Edward Celfyn Thomas. Among those who have provided supportive voices are Dr Lynne Gornall, Lucy Sweetman and Dr Lyndon Murphy. To all those who have taken interest in the writing of this book, we are most grateful.

1 Introduction

1.1 Key aspects of the study

This book is primarily about using the example of the float glass process as an illustration of entrepreneurship, business process innovation and diffusion. Innovation and diffusion theory are front and centre, with specific industrial history forming a detailed case study. The importance of the glass industry as a pertinent example for study of innovation activities has been recognised for some time (Jewkes et al., 1969). It is on this basis that this story is a suitable case study in this instance.

Considerable material has been published about the float glass process in terms of its history (Barker, 1977, 1994), technical features (Pilkington, 1969), shop floor viewpoint (Grundy, 1990), development (Bricknell, 2009), innovation (Uusitalo, 2000, 2014), and family history (Pilkington, 2010). But little has been written, resulting in a "gap" in the literature, in terms of the management of the early float glass start-ups. This case study seeks to remedy this by recounting developments through the "lens" of one of the leading glass technologists involved in the process at the time, using historical and archival material, and artefacts from the period.

The key point of the book is that it is believed by "telling" the story of the early float glass start-ups, important aspects of the business of new process diffusion involved in the management of early start-ups will come to light. This will be in terms of the case study concerning business origins of the process and its invention, progressing to innovation, competition in the market, first successful production, licensing and patents, and the management of the start-ups leading to market leadership.

A book concerning the early float glass start-ups topic is relevant to the contemporary literature in the area of innovation due to being a concise case study of significance to technology, entrepreneurship and innovation. In recent years there has been considerable growth and interest in entrepreneurship and innovation as an important study area at leading universities.

The combination of the foundation of the theoretical study of innovation, diffusion and start-ups with an understanding of the application of this to new process diffusion provides a powerful and rich source of material for academic study. This is an ideal way to develop the content to make the most of this subject. The study will appeal to the book's intended readership by providing a convenient and interesting way for undergraduate and post-graduate students and postdoctoral researchers to gain an understanding of the processes involved in new process diffusion and start-up management.

1.2   Interrelationship between invention, innovation and business development

Innovation can be defined as either the "application of a new method or device" (Collins, 1997) or the "successful exploitation" of a new idea (Thomas and Rhisiart, 2000). According to Baregheh et al. (2009), innovation is "the multi-stage process whereby organisations transform ideas into new/improved products, services or processes, in order to advance, compete and differentiate themselves successfully in their marketplace". In terms of successful innovation Mariello (2007) identifies five discrete and essential stages: (1) idea generation and mobilisation – the generation stage is the starting line for new ideas; (2) advocacy and screening; (3) experimentation; (4) commercialisation; and (5) diffusion and implementation.

Innovation literature has long demonstrated the importance of external sources in the development of successful innovation (Carter and Williams, 1957). These studies tended to focus on the identification of the sources and types of knowledge and technology, often neglecting the nature and origins of the relationship linking the recipient (the innovator) to the source of technological innovation. In innovation support networks technology equates with knowledge. Within university-industry link systems a multiplicity of technology transfer mechanisms are apparent, which appear to be well integrated (Cheese, 1993). Internal research and development (R&D) not only produces new information but also evolves external know-how and technology (Cohen and Levinthal, 1989). Freeman (1991) has argued that "the successful exploitation of imported technology is strongly related to the capacity to adapt and improve this technology through indigenous R&D".

Much has been written about invention and inventive activity. Published work typically describes inventive activity on a historical-developmental basis or as a collection of case studies, presenting qualitative findings in relation to the inventive developments taking place. Indeed, the relationship between invention and innovation has involved much discussion. Innovation is defined by Kanter (1983) as involving "creative use as well as original invention", and simply it is defined by Mellor (2005) as "creativity plus application" or "invention plus application". Burns (2007) further reports that "invention is the extreme and riskiest form of innovation". In particular, Bolton and Thompson (2000) highlight creativity in the invention and innovation process.

The interrelationship between invention and innovation is both of theoretical and practical significance. It may involve inventors in all aspects of the process of product, process or service development, but also it can involve them separately. The latter case is exemplified historically by Adam Smith (1776), who observed that "all the improvements in machinery, however, have by no means been the inventions of those who had occasion to use the machines". He also considered the way in which the division of labour promoted specialised inventions. This is articulated by Marx (1858), who notes "invention then becomes a branch of business, and the application of science to immediate production aims at determining the inventions at the same time as it solicits them". Freeman and Soete (1997, p. 15) develop this theme of invention as "an essential condition of economic progress and a critical element in the competitive struggle of enterprises and of nation-states".

Freeman and Soete (1997, p. 16) remark that "although most economists have made a deferential nod in the direction of technological change, few have stopped to examine it". This paradox has been explained by Jewkes et al. (1969) in terms of the ignorance of science and technology by economists, their pre-occupation with the trade cycle and employment problems, and limited statistics. This was demonstrated by Jewkes et al. (1969) in their study of "The Sources of Invention" and has been confirmed before and since by empirical studies. Freeman and Soete (1997, p. 17) develop this argument regarding the neglect of invention, since it

> was not only due to other pre-occupations of economists nor to their ignorance of technology; they were also the victims of their own assumptions and commitment to accepted systems of thought. These tended to treat the flow of new knowledge, of inventions. . . . as outside the framework of economic models, or more strictly, as "exogenous variables".

The distinction between invention and innovation was originally owed to Schumpeter (1934, 1961) and has since become part of economic theory. Freeman and Soete (1997, p. 22) add, "an invention is an idea, a sketch or a model for a new improved device, product, process or system. Such inventions may often (not always) be patented but they do not necessarily lead to technical innovations". Also, "the chain of events from invention or specification to social application is often longer and hazardous" (Freeman

Table 1.1 Inputs and outputs of inventive work

Process	Inventive inputs		Inventive outputs	
	Inputs from	Other inputs	Output	Other outputs
Inventive work	Orders from entrepreneurs Inventive work development	Outputs of research	New technological problems Unexplainable successes and failures	Patents Non-patentable inventions

Source: Adapted from Ames (1961) and Freeman and Soete (1997).

and Soete, 1997, p. 22). The crucial role of the entrepreneur in this complex process was recognised by Schumpeter (1934, 1961), although he did not consider the study of invention to be of significance in itself. According to Johnson (2018, p. 8) "entrepreneurs spot opportunities and seize them with vigour. They are energetic, ambitious and competitive". A summary of the inputs and outputs of the inventive process, based on Ames (1961) and Freeman and Soete (1997), is presented in Table 1.1.

In the twentieth century, according to Freeman and Soete (1997), there was a shift towards large-scale corporate R&D. This is contrary to the interpretation provided by Jewkes et al. (1969) in their classic study *The Sources of Invention*, as already mentioned. In this, they reduce the difference between the nineteenth and twentieth centuries and minimise the importance of corporate R&D. Moreover, they argued that important twentieth-century inventions were the result of individual inventors similar to the nineteenth century. In fact, they concede that due to the extortionate development costs, large-scale corporations will often still be necessary to bring inventions into commercial exploitation. Indeed, out of 64 major twentieth-century inventions, 40 were attributed to individual inventors compared to 24 from corporate R&D, and out of those 40, half were dependent for commercial development on large firms. One of the twentieth-century inventions they considered was the float glass process.

Freeman and Soete (1997) maintained from the standpoint of economics that it was innovation that was of central interest rather than invention, although they did not deny the importance of invention or the vital contribution creative individuals make to invention. This has been highlighted by Johnson (1975), who recognised the economic significance of invention itself in terms of its process and relationship, to the size of the firm and the role of the individual inventor. But they do note, even on Jewkes et al.'s account of major inventions, that there has been a shift since the early twentieth century to a larger contribution from inventors associated with corporate R&D.

According to Freeman and Soete (1997, p. 169):

> the test of successful entrepreneurship and good management is the capacity to link together. . . . technical and market possibilities. . . . Innovation is a coupling process and the coupling first takes place in the minds of imaginative people. . . . But once the idea has "clicked" in the mind of the inventor or entrepreneur, there is still a long way to go before it becomes a successful innovation. . . . The one-(person) inventor-entrepreneur. . . . may very much simplify this process in the early stages of a new innovating firm, but in the later stages and in any established firm the "coupling" process involves linking and co-ordinating different sections, departments and individuals.

A fundamental question regarding the role of the inventor is whether invention depends on inventors in terms of national and regional policies, which may aim to liberate individual "inventiveness". According to Norris and Vaizey (1973), this widely held view may be false. It is debatable whether this is the case since although most inventions are promulgated by individuals, due to a creative idea emerging from one person, it is possible for two or more people to get together to formulate an idea. This is contrary to Norris and Vaizey's assertion that "groups of people do not tend to produce creative ideas" (Norris and Vaizey, 1973, p. 36). This leads to the possibility of co-invention, and this is supported by the research reported by Thomas et al. (2009) in a survey of inventors which provides evidence of inventors working together in a number of cases.

Ideas developed at any time may be linked or they may be separate. The corporate or institutional inventor may be a core tenured employee who is working in a specific area the results of which will be retained by the employer. The relationships between invention, innovation and entrepreneurship, inventors, innovators and entrepreneurs, and micro, small and medium-sized enterprises (SMEs) and large corporations are illustrated in Table 1.2.

Table 1.2 The interrelationships between invention, innovation and entrepreneurship

Activity/Level	Invention	Innovation	Entrepreneurship
Micro	Individual/lone inventor	Innovator	Entrepreneur
Small and medium-sized enterprise	Company inventor	Innovation champions	Entrepreneur/ intrapreneur
Large company or organisation	Institutional/ corporate inventor	Project champions	Intrapreneur

Source: Thomas and Gornall (2002).

The measurement of the relative magnitude of inventive activity by inventors is problematic due to the absence of expenditure on this type of activity. As a consequence, measurement is currently based purely on outcomes. The two main sources of information are therefore patent statistics and information on significant inventions. According to Kuznets (1962), there are four possible dimensions to an invention: a technical and an economic magnitude, and a past and a future. The technical past relates to the magnitude of the technical problem resolved by the invention. Consequently, some inventions are of a greater magnitude than others. The technical future can be measured according to the size of the invention, which is dependent on the inventions that follow. An example of this is a number of inventions and innovations that were related to, and followed, the float glass process. The economic past of an invention involves the cost and is measured according to the resources used. Lastly, the economic future of an invention involves the production of new goods or services and can enable cost reductions. Although the above measures act as a conceptual framework, it remains an educated guess to determine the difference between significant and insignificant inventions. Jewkes et al. (1969), in their work on the most important inventions in the twentieth century, as already described, assembled a list in their judgement of the most significant inventions including the float glass process.

Factors affecting the inventor as a major source of invention include time, "atmosphere", finance and technological resources. The complexities of finding finance by an inventor are explored by Hobbs (2006) in terms of the inventor-investor relationship. With regard to time, businesses will be interested in inventions that will yield a pay-off within a short period of time, and many firms will expect expenditures to be paid off within five years (Norris and Vaizey, 1973; Freeman and Soete, 1997). Because five years will have to include the process of recouping spending on research, invention, innovation and marketing, this will restrict the magnitude of the scale of the advancement of knowledge. As a consequence, most company R&D is concerned with small improvements.

A factor working against invention is the problem of providing the right "atmosphere". Another major factor working against the individual inventor is the lack of finance, and this is why they appear to have declined in importance in the twentieth century. Much invention will also require specialised technological equipment with a cost beyond the reach of many individual inventors. It could therefore be expected that the role of the individual inventor would be most significant in areas which do not need large amounts of expensive technological equipment. Norris and Vaizey (1973) state that since inventions can be a result of many highly trained personnel working methodically on problems with considerable financial backing, it is clearly the case that there can be both contentions that inventions have been the result of both team and individual work. They therefore surmise that the individual inventor will continue to play a significant role.

According to Spence (1995), innovation is often used to indicate something new, created or produced, and it is commonly confused with invention. Whereas inventions can be seen as innovations because they are new, innovations are not necessarily inventions. Spence (1995) further says that innovations may be long-established ideas, products or services involving a new application and consequently may be considered novel. An interesting development of the classic distinction between innovation and invention is with regard to technical novelties (McKelvey, 1997). These may be hidden in an inventor's garage or in an R&D department. They may also be mentioned in patents but remain unused, developed or sold and are therefore technical inventions. As technical novelties, they include a combination of techniques, knowledge and technologies. In fact, inventions become innovations when they are used for marketable products or sold. Indeed, many innovations will have a degree of technical novelty and involve interaction with the marketplace.

"Collective invention" "is the free exchange of information about new techniques and plant designs among actual and potential competitors" (Foray, 1997). This has been described in the case of the iron industry:

> If a firm constructed a new plant of novel design and that plant proved to have lower costs than other plants, these facts were made available to other firms in the industry and to potential entrants. The next firm constructing a new plant could build on the experience of the first by introducing and extending the design change that had proved profitable. The operating characteristics of this second plant would then also be made available to potential investors. In this way fruitful lines of technical advance were identified and pursued.
>
> (Allen, 1983, p. 2)

It is through this behaviour that cumulative advance takes place (Ehrnberg and Jacobsson, 1997). It appears that individual invention has become less important and collective invention more important (Edquist and Johnson, 1997). Moreover, innovation is the interaction of an invention into a use that has economic value. Inventors will design and develop new products and services and innovators will recognise the opportunities (Burns, 2007), take the risk and accept the challenges. It should also be remembered that inventions solve problems and will lead to other inventions.

1.3 Importance of research and development

Two roles for R&D are suggested by Griffith et al. (2004): to stimulate innovation and to create an understanding of discoveries by others which to the originating firm are confidential. The Schumpeterian hypothesis (1934, 1942) suggests that market concentration and large production units for R&D-intensive industries are not necessarily confirmed through empirical evidence. The process of "creative destruction" (Schumpeter, 1934, 1942) means that firms in technology dynamic industries, where there is oligopolistic competition, will need to innovate to maintain their position in the market. Caballero and Jaffe (1993) have provided empirical support for this hypothesis, and according to Nelson (1990) the views of R&D and company managers also support this point. Levin et al. (1987), in a survey of large corporations in the United States, examined a number of methods used by firms to protect the competitive advantage of new or improved processes and products; these were patents, secrecy, lead time, moving quickly along the learning curve and sales and service. With "first mover advantage", it was found that secrecy was the most widely used method to protect intellectual property (IP) in industry (Arundel, 2001).

Once knowledge is created, due to non-exclusion it is hard to stop others using it and to keep it private, and this is the non-appropriation problem (Revesz and Boldeman, 2006). In relation to this, Quah (2003) has considered with regard to the information society the public good aspects. Further to this, with knowledge there is the implication of only charging for marginal dissemination costs (Arrow, 1962). As a result additional learning costs will be incurred by the user when making use of this knowledge (Mandeville et al., 1982). Whereas scientific knowledge, which contributes to greater understanding instead of new applications in the public domain, is more available, know-how and technical information ("proprietary" knowledge) tends not to be publicised, and surveys of R&D and business managers have supported the view that patent disclosures and technical publications do not play a significant role in the provision of technology information to innovative firms (Revesz and Boldeman, 2006). Indeed, a survey in the United States by Schuchman (1981) found that engineers involved with new technologies relied on in-house expertise and talking to colleagues for information that was relevant and that they tended not to use technical publications. Books used when developing the float glass process included works by Stanworth (1950) and Jones (1956) (see Appendix 3).

A number of surveys have been undertaken to consider the time delay and cost in the imitation of inventions (Revesz and Boldeman, 2006). For example, more than 120 respondents to a survey (mostly United States

R&D executives) were asked by Levin et al. (1987) for an estimation of time and costs needed to copy innovations by a competitor, and it was found that in less than five years most inventions could be imitated. Similarly, Mansfield et al. (1981), and Mansfield (1985) revealed that reverse engineering, personal contacts and the movement of staff between companies were the principal sources of the leakages of information. According to Griffith et al. (2004), two roles for R&D are those of (1) stimulating innovation and (2) enabling understanding and the imitation of discoveries which remain confidential by other originating firms. With the float glass process, R&D played an important role. R&D therefore plays an important role for the development of an "absorptive capacity" and is equally critical for technology transfer and innovation (Revesz and Boldeman, 2006). R&D appears to stimulate innovation indirectly by technology transfer or directly by those involved with leading edge technology frontiers (Revesz and Boldeman, 2006).

Although there appears to be no data on the commercial return from R&D activities, case studies of firm managers show that they will invest in R&D due to a competitor's technology advances and the fear of being out of business (Revesz and Boldeman, 2006). In a study by Revesz and Lattimore (2001), no statistical positive significance between R&D intensity and firm profitability was found, and a survey by Jaruzelski et al. (2005) also found no direct relationship between R&D spending and corporate success. The definition of R&D by the OECD is:

> Research and experimental development (R&D) comprise creative work undertaken on a systematic basis in order to increase the stock of knowledge, including knowledge of man, culture and society, and the use of this stock of knowledge to devise new applications.
>
> (OECD, 2002)

Unfortunately, a major drawback of case studies is that they only consider innovations that are successful (Revesz and Boldeman, 2006). Alternatively, case studies can be useful when information about R&D costs and outcomes, which are commercially sensitive, is available from private businesses. Many R&D studies have only considered manufacturing since it represents the largest spend on R&D than any other sector (Revesz and Boldeman, 2006). The cost savings for 12 manufacturing sectors in the United States were estimated by Nadiri and Theofanis (1994) – the social manufacturing rate of return on public R&D was found to be between 6% and 9% by adding the marginal cost savings estimates. The rate at which companies registered significant product innovations and patents across technology fields in the United States was analysed by Acs et al. (1994), who found that own R&D activity was important for large businesses who ran their own laboratories. R&D may well influence considerable knowledge spillovers to business through "tacit" knowledge, training of researchers, collaborative ventures, resolving technological dilemmas and scientific and new discoveries (Revesz and Boldeman, 2006).

With regard to demand, it is apparent that the motivation to undertake R&D has involved variables representing market demand conditions which present demand as a major influence on such decisions (Crespi et al., 2003). Unfortunately, as noted by Mowery and Rosenberg (1979) this does not convey much since managers will consider the demand outcome before undertaking the development process, which is likely to be expensive. According to the Schumpeterian perspective, innovation and R&D activities in modern times have required large firms or concentrated industries (Crespi et al., 2003). This was the case with the float glass process. Consequently, there will be sectors where the spend on R&D will be determined by the minimum operation scale (Acs and Audretsch (1990), Audretsch (1995) and Audretsch and Vivarelli (1996) explain this according to different technological regimes across the different sectors and firm size). Cohen (1995) notes that the scale economies in R&D may be a possible explanation for the impact of large-sized firms.

According to von Tunzelmann (1995), all productive units involve the four functions of administration and finance, products, production processes and technology (with augmentation by R&D). In the literature on scale economies in R&D there is justification for merging large high-technology firms (HTFs) (Fisher and Temin, 1973; Kohn and Scott, 1982). The cycle time is the speed for R&D to be turned into new products, and in order to be first to market there will be pressure for businesses to shorten the time (Crespi et al., 2003). Setting aside increase in complexity, a faster cycle time has its own costs (Scherer and Ross, 1990). Within firms there is a danger that there will be too narrow a focus on innovation and R&D, since as well as the ability to create new products and processes absorptive capacity will depend on the other resources and functions within and outside the organisation (Crespi et al., 2003). Teece (1986) has called these other resources "complementary assets".

According to many studies a significant determinant of R&D appears to be financing of innovation and the role of cash flow (Crespi et al., 2003). In the literature on appropriate methods for the evaluation of the financing of R&D, Myers (1984) has suggested options valuations instead of payback procedures or conventional discounted cash flow (DCF). A problem is that if a company leaves an R&D project, it may be far more expensive to return at a later date (Mitchell and Hamilton, 1988). As noted by Thomas et al. (2018, p. 222), "matters can also shift quite significantly during the course of an association or project lifecycle, resetting the position of participants, even leaders". Here "leadership is a crucial variable too, in setting the ethos of and for the relationship" (Thomas et al., 2018, p. 224). Marketing functions also need to be taken into account since there may be a considerable gulf between marketing and R&D (Crespi et al., 2003).

1.4 Technological development of firms through new process diffusion

Due to the increasing influence of technology on company strategy and the important role of technological progress in the stimulation of industrial development, and the complexity and diversity of modern technological practices (Gold, 1987), many businesses experience difficulty in gaining access to certain technologies. Indeed, it is increasingly suggested that access to technologies by businesses can best be achieved by encouraging the formation of networks of innovators. Such collaborative arrangements are essential to improving the competitive position of companies, predominantly through the accomplishment of mutually beneficial goals such as the acquisition of state-of-the-art technology (Forrest and Martin, 1992). Such innovation support networks serve to externalise the innovation function through the transfer of technology between firms (Lawton-Smith et al., 1991). During the last 25 years industrial innovation has become significantly more of a networking process, with collaborations increasing considerably (Aldrich and Sasaki, 1995). Indeed there is mounting evidence of network relationships between businesses, especially the transfer of technology (Lipparini and Sobrero, 1994). It is likely that businesses will become more dependent on external sources during the innovation process. In this study, with regard to the innovation process we are particularly interested in process innovation, and especially new process diffusion. According to InnoviSCOP (2018), "process innovation means the implementation of a new or significantly improved production or delivery method (including significant changes in techniques, equipment and/or software)". Also, process innovation is a new or significantly improved way of doing things in a business that typically increases production levels and decreases costs (Study.com, 2018).

1.5 Structure of the book

The book contains four chapters: Chapter 1, Introduction; Chapter 2, New Process Diffusion; Chapter 3, Case study – Management of the Early Float Glass Start-Ups; and Chapter 4, Conclusions. These are described below.

Chapter 1: introduction

The introductory chapter has introduced new process diffusion by considering innovation as a "multi-stage process whereby organisations transform ideas into new/improved products, services or processes, in order to advance, compete and differentiate themselves successfully in their marketplace" (Baregheh et al., 2009). It has presented the distinction between invention and innovation and the interrelationships between invention, innovation and business development (Thomas et al., 2009). It has also investigated the importance of R&D in terms of the technological development of firms.

Chapter 2: new process diffusion

The second chapter considers new process diffusion in the form of new or improved technology, and the transmission of knowledge or technical expertise (Thomas, 1999, 2000). This is investigated in terms of technology diffusion, technology transfer, a model of technology diffusion (Thomas et al., 2001, 2004, 2006) and the importance of these activities for new process diffusion and business partnerships (Thomas et al., 2013).

Chapter 3: case study – management of the early float glass start-ups

The third chapter, which concerns the case study of the management of the early float glass start-ups, includes sections on the background of the case study, process development, process diffusion, managing the early start-ups and an overview.

Chapter 4: conclusions

The book concludes by reflecting on the example of the float glass process as an illustration of entrepreneurship, business process innovation and diffusion. Through innovation and diffusion theory being investigated, with specific industrial history forming a detailed case study of the early float glass start-ups, the importance of successful start-ups to new process diffusion will be emphasised.

The next chapter provides an overview of new process diffusion and start-ups involving technology diffusion and transfer with relevance to the management of the early float glass start-ups.

References

Acs, Z. J. and Audretsch, D. B. (1990) *Innovation and Small Firms*, Cambridge, MA: MIT Press.

Acs, Z. J., Audretsch, D. B. and Feldman, M. P. (1994) R&D spillovers and innovative activity, *Managerial and Decision Economics*, 15(2), March, 131–138.

Aldrich, H. E. and Sasaki, T. (1995) R&D consortia in the United States and Japan, *Research Policy*, 24(2), 301–316.

Allen, R. C. (1983) Collective invention, *Journal of Economic Behaviour and Economic Organization*, 4, 1–24.

Ames, E. (1961) Research, invention, development and innovation, *American Economic Review*, 51(3), 370–381.

Arrow, K. J. (1962) Economic welfare and the allocation of resources for invention, in *The Rate and Direction of Inventive Activity*, Princeton, NJ: Princeton University Press, National Bureau of Economic Research.

Arundel, A. (2001) The relative effectiveness of patents and secrecy for appropriation, *Research Policy*, 30, 611–624.

Audretsch, D. B. (1995) *Innovation and Industry Evolution*, Cambridge, MA: MIT Press.

Audretsch, D. B. and Vivarelli, M. (1996) Firm's size and R&D spillovers: Evidence from Italy, *Small Business Economics*, 8(3), June, 249–258.

Baregheh, A., Rowley, J. and Sambrook, S. (2009) Towards a multidisciplinary definition of innovation, *Management Decision*, 47(8), 1323–1339.

Barker, T. C. (1977) *The Glassmakers: Pilkington: The Rise of an International Company 1826–1976*, London: Weidenfeld and Nicolson.

Barker, T. C. (1994) *Pilkington: An Age of Glass: The Illustrated History*, London: Boxtree.

Bolton, B. and Thompson, J. (2000) *Entrepreneurs: Talent, Temperament, Technique*, Oxford: Butterworth-Heinemann.

Bricknell, D. J. (2009) *Float: Pilkingtons' Glass Revolution*, Lancaster: Crucible Books.

Burns, P. (2007) *Entrepreneurship and Small Business*, Basingstoke: Palgrave Macmillan.

Caballero, R. J. and Jaffe, A. B. (1993) *How High Are the Giants' Shoulders: An Empirical Assessment of Knowledge Spillovers and Creative Destruction in a Model of Economic Growth*, National Bureau of Economic Research, Working Paper No. 4370, Cambridge, MA.

Carter, C. and Williams, B. (1957) *Industry and Technical Progress: Factors Affecting the Speed and Application of Science*, London: Oxford University Press.

Cheese, J. (1993) Sourcing technology: Industry and higher education in Germany and the UK, *Industry and Higher Education*, March, 30–38.

Cohen, W. M. (1995) Empirical studies of innovative activity, in P. Stoneman (ed.), *Handbook of the Economics of Innovation and Technical Change*, Oxford: Blackwell.

Cohen, W. M. and Levinthal, D. (1989) Innovation and learning: The two faces of R&D, *Economic Journal*, 99, 569–596.

Collins (1997) *Collins Concise Dictionary*, Glasgow: HarperCollins.

Crespi, G., Patel, P. and von Tunzelmann, N. (2003) *Literature Survey on Business Attitudes to R&D*, Science Policy Research Unit (SPRU), Brighton: University of Sussex.

Edquist, C. and Johnson, B. (1997) Institutions and organizations in systems of innovation, in C. Edquist (ed.), *Systems of Innovation: Technologies, Institutions and Organizations*, London: Pinter, p. 53.

Ehrnberg, E. and Jacobsson, S. (1997) Technological discontinuities and incumbents' performance: An analytical framework, in C. Edquist (ed.), *Systems of Innovation: Technologies, Institutions and Organizations*, London: Pinter, pp. 318–341.

Fisher, F. M. and Temin, P. (1973) Returns to scale in research and development: What does the Schumpeterian hypothesis imply? *Journal of Political Economy*, 81, 56–70.

Foray, D. (1997) Generation and distribution of technological knowledge: Incentives, norms, and institutions, in C. Edquist (ed.), *Systems of Innovation: Technologies, Institutions and Organizations*, London: Pinter, p. 73.

Forrest, J. E. and Martin, M.J.C. (1992) Strategic alliances between large and small research intensive organisations: Experiences in the biotechnology industry, *R&D Management*, 22, 41–67.

Freeman, C. (1991) Networks of innovators: A synthesis of research issues, *Research Policy*, 20(5), 499–514.

Freeman, C. and Soete, L. (1997) *The economics of industrial innovation*, 3rd edn, London: Pinter.

Gold, B. (1987) Technological innovation and economic performance, *Omega*, 15(5), 361–370.

Griffith, R., Redding, S. and van Reenen, J. (2004) Mapping the two faces of R&D: Productivity growth in a panel of OECD Industries, *The Review of Economics and Statistics*, 86(4), 882–895.

Grundy, T. (1990) *The Global Miracle of Float Glass: A Tribute to St Helens and Its Glass Workers*, St Helens: Chalon Press.

Hobbs, F. (2006) The inventor-investor conundrum, *Industry and Higher Education*, 20(6), December, 381–385.

InnoviSCOP (2018) Process Innovation-Definition-InnoviSCOP, www.innoviscop.com/en/definitions/process-innovation.

Jaruzelski, B., Dehoff, K. and Bordia, R. (2005) The Booz Allen Hamilton global innovation 1000, *Strategy and Business*, 41, Winter.

Jewkes, J., Sawers, D. and Stillerman, R. (1969) *The Sources of Invention*, 2nd edn, London: Macmillan.

Johnson, L. (2018) What it takes to a successful entrepreneur: The rich list 2018, *Sunday Times Magazine*, May 13, p. 8.

Johnson, P. S. (1975) *The Economics of Invention and Innovation*, London: Martin Robertson, pp. 29–50, 51–71 and 244–250.

Jones, G. O. (1956) *Glass*, London and New York: Methuen and John Wiley.

Kanter, R. M. (1983) *The Change Masters: Innovation and Productivity in American Corporations*, New York: Simon and Schuster.

Kohn, M. and Scott, T. J. (1982) Scale economies in research and development, *Journal of Industrial Economics*, 30, 239–250.

Kuznets, S. (1962) Inventive activity: Problems of definition and measurement, *National Bureau Committee for Economic Research, The Rate and Direction of Inventive Activity*, Princeton University Press, Princeton.

Lawton-Smith, H., Dickson, K. and Lloyd-Smith, S. (1991) There are two sides to every story: Innovation and collaboration within networks of large and small firms, *Research Policy*, 20, 457–468.

Levin, R., Klevorick, A. K., Nelson, R. and Winter, S. G. (1987) *Appropriating the Returns from Industrial Research and Development*, Brookings Papers on Economic Activity, vol. 3, Washington, DC.

Lipparini, A. and Sobrero, M. (1994) The glue and the pieces: Entrepreneurship and innovation in small firm networks, *Journal of Business Venturing*, 9, 125–140.

Mandeville, T. D., Lamberton, D. M. and Bishop, E. J. (1982) *Economic Effects of the Australian Patent System*, Canberra: AGPS.

Mansfield, E. (1985) How rapidly does new industrial technology leak out? *The Journal of Industrial Economics*, 34(2), December, 217–223.

Mansfield, E., Schwartz, M. and Wagner, S. (1981) Imitation costs and patents: An empirical study, *Economic Journal*, 91(364), 907–918.

Mariello, A. (2007) The five stages of successful innovation, *Opinion and Analysis*, Spring, April 1, https://sloanreview.mit.edu/article/the-five-stages-of-successful-innovation/.

Marx, K. (1858) *Grundrisse*, London: Allen Lane Edn, 1973.

McKelvey, M. (1997) Using evolutionary theory to define systems of innovation, in C. Edquist (ed.), *Systems of Innovation: Technologies, Institutions and Organizations*, London: Pinter, p. 201.

Mellor, R. B. (2005) *Sources and Spread of Innovation in Small e-Commerce Companies*, Skodsborgvej, Denmark: Forlaget Globe.

Mitchell, G. and Hamilton, W. (1988) Managing R&D as a strategic option, *Research Technology Management*, 31(3), 15–22.

Mowery, D. C. and Rosenberg, N. (1979) The influence of market demand upon innovation: A critical review of some recent empirical studies, *Research Policy*, 8, 103–153.

Myers, S. (1984) Finance theory and finance strategy, *Interfaces*, 14, 126–137.

Nadiri, M. I. and Theofanis, P. M. (1994) The effects of public infrastructure and R&D capital on the cost structure and performance of U.S. manufacturing industries, *Review of Economics and Statistics*, 76(1), February, 22–37.

Nelson, R. R. (1990) Capitalism as an engine of progress, *Research Policy*, 19.

Norris, K. and Vaizey, J. (1973) *The Economics of Research and Technology*, London: George Allen & Unwin, pp. 36–42.

Organisation for Economic Co-Operation and Development (OECD) (2002) *Frascati Manual*, Paris: OECD.

Pilkington, L.A.B. (1969) Review lecture: The float glass process, *Proceedings of the Royal Society of London A*, 314, 1–25.

Pilkington, D. (2010) *A Glass Act*, Newcastle upon Tyne: Tyneside Free Press.

Quah, D. (2003) Digital goods and the new economy, in D. C. Jones (ed.), *New Economy Handbook*, San Diego: Elsevier Academic Press, pp. 291–323.

Revesz, J. and Boldeman, L. (2006) *The Economic Impact of ICT R&D: A Literature Review and Some Australian Estimates*, Occasional Economic Paper, Australian Government Department of Communications, Information Technology and the Arts, Commonwealth of Australia, November, pp. 1–140.

Revesz, J. and Lattimore, R. (2001) *Statistical Analysis of the Use and Impact of Government Business Programmes*, Staff Research Paper, Productivity Commission, Canberra.

Scherer, F. M. and Ross, D. (1990) *Industrial Market Structure and Economic Performance*, 3rd edn, Boston: Houghton Mifflin.

Schuchman, H. (1981) *Information transfer in engineering*, Washington, DC: Futures Group.

Schumpeter, J. (1934) *The Theory of Economic Development*, Cambridge, MA: Harvard University Press.

Schumpeter, J. (1942) *Capitalism, Socialism and Democracy*, New York: Harper.

Schumpeter, J. (1961) *The Theory of Economic Development*, Oxford: Oxford University Press, 1934.

Smith, A. (1776) *An Inquiry into the Nature and Causes of the Wealth of Nations*, Dent edn 1910, London, p. 8.

Spence, W. R. (1995) *Innovation: The Communication of Change in Ideas, Practices and Products*, London: Chapman & Hall, p. 4.

Stanworth, J. E. (1950) *Physical Properties of Glass*, Oxford: Oxford University Press.

Study.com (2018) Process Innovation: Types & Examples-Video & Lesson Transcript, www.study.com/academy/lesson/process-innovation-types-examples.html.

Teece, D. J. (1986) Profiting from technological innovation: Implications for integration, collaboration, licensing and public policy, *Research Policy*, 15, 285–305.

Thomas, B. (1999) A model of the diffusion of technology into SMEs, *Proceedings of the 44th International Council for Small Business (ICSB) World Conference: Innovation and Economic Development*, Naples, June 20–23.

Thomas, B. (2000) *Triple Entrepreneurial Connection: Colleges, Government and Industry*, London: Janus.

Thomas, B. and Gornall, L. (2002) The role of the individual inventor and the implications for innovation and entrepreneurship: A view from Wales, *25th ISBA National Small Firms Policy and Research Conference: Competing Perspectives of Small Business and Entrepreneurship*, Brighton, November 13–15.

Thomas, B., Gornall, L., and Murphy, L. (2018) Valuing European partnerships: Memories of cross-national leadership in UK higher education projects, in L. Gornall, B. Thomas and L. Sweetman (eds.), *Exploring Consensual Leadership in Higher Education: Co-Operation, Collaboration and Partnership*, London: Bloomsbury, pp. 223–241.

Thomas, B., Gornall, L., Packham, G. and Miller, C. (2009) The individual inventor and the implications for innovation and entrepreneurship, *Industry and Higher Education*, 23(5), 391–403.

Thomas, B., Murphy, L. and Lewis, A. (2013) The management of university partnerships in the UK with special reference to Wales, *ICBR Journal*, 2(1), 19–41.

Thomas, B., Packham, G. and Miller, C. (2001) A temporal model of technology diffusion into small firms in Wales, *Industry and Higher Education*, August, 279–288.

Thomas, B., Packham, G. and Miller, C. (2006) Technological innovation, entrepreneurship, higher education and economic regeneration in Wales: A policy study, *Industry and Higher Education*, 20(6), December, 433–440.

Thomas, B., Packham, G., Miller, C. and Brooksbank, D. (2004) The use of web sites for SME innovation and technology support services in Wales, *Journal of Small Business and Enterprise Development*, 11(3), 400–407.

Thomas, M. and Rhisiart, M. (2000) Innovative Wales, in J. Bryan and C. Jones (eds.), *Wales in the 21st Century: An Economic Future*, London: Macmillan Business, pp. 115–122.

Uusitalo, O. (2000) Impact of technological changes on the industry: The case of the Scandinavian flat glass industry in 1910–1990, *Schumpeter 2000*, 1–19.

Uusitalo, O. (2014) *Float Glass Innovation in the Flat Glass Industry*, London: Springer.

Von Tunzelmann, G. N. (1995) *Technology and Industrial Progress: The Foundations of Economic Growth*, Aldershot: Edward Elgar.

2 New process diffusion

2.1 Overview of new process diffusion and start-ups

New process diffusion is important for the development of manufacturing processes which are defined as

> the steps through which raw materials are transformed into a final product. The manufacturing process begins with the creation of the materials from which the design is made. These materials are then modified through manufacturing processes to become the required part.
>
> (Chegg, 2018)

Further to this, new processes involve process innovation, which can be defined as follows:

> process innovation means the implementation of a new or significantly improved production or delivery method (including significant changes in techniques, equipment and/or software). Minor changes or improvements, an increase in production or service capabilities through the addition of manufacturing or logistical systems which are very similar to those already in use, ceasing to use a process, simple capital replacement or extension, changes resulting purely from changes in factor prices, customisation, regular seasonal and other cyclical changes, trading of new or significantly improved products are not considered innovations.
>
> (InnoviSCOP, 2018)

Since this study investigates the early float glass start-ups which were undertaken on a global basis, it is appropriate to take into consideration the characteristics of global start-ups. Critical success factors (CSFs) for global start-ups (not in rank order) have been reported as:

> managerial global vision from inception; high degree of previous international experience on behalf of managers; management commitment; strong use of personal and business networks (networking); market knowledge and market commitment; unique intangible assets based on knowledge management; high value creation through product differentiation, leading edge technology products, technological innovativeness (usually associated with a greater use of IT), and quality leadership; niche focussed, proactive international strategy in geographically spread lead markets around the World from the very beginning; narrowly defined customer groups with strong customer orientation and close customer relationships; flexibility to adapt to rapidly changing external conditions and circumstances.
>
> (Rialp-Criado et al., 2002, pp. 25–26)

According to Wakkee et al. (2003), from their discussion of the literature on the definition of a global start-up, five relevant characteristics are apparent: "(1) the diversity or scope of the international activities; (2) the company age; (3) the timing of international activities (time to entry); (4) the global diversity of the international activities; (5) the purpose of the international activities (strategic choice)" (Wakkee et al., 2003, p. 13).

Here "'opportunity recognition' seeks to pursue opportunities wherever they arise (i.e. global or in an unlimited number of countries around the world); it co-ordinates multiple activities in the value chain through the interaction with network actors around the World. The entrepreneur(ial) team) leading the firm is internationally experienced and skilled". (Wakkee et al., 2003, p. 14).

2.2 Technology diffusion

Three aspects of the diffusion of technology considered in this chapter are, first, the investigation of technology diffusion (Brooksbank et al., 2001) in the form of new or improved technology; second, mechanisms involved in the transfer of technology into the innovative firm, and the development of a model of technology diffusion including external sources, and channels of technology transfer; and third, to relate the model to new process diffusion and business partnerships and the early float glass start-ups.

All models of technology diffusion, including refined models such as the Bass Norton model, are a simplification of reality (Islam and Meade, 1997). One theoretical model that has informed policies is the centre-periphery model (Schon, 1971) which rests on three basic assumptions:

1   The technology to be diffused exists prior to its diffusion;
2   Technology diffusion takes place from the source outwards to firms; and
3   The support of technology diffusion involves incentives, provision of resources and training.

Diffusion will take place from the source of the technology through channels by a "diffuser", using a transfer mechanism, to the firm. The effectiveness of the system will depend upon the resources available to the external source to enable the transfer, the efficiency of the diffuser and the mechanism involved, and the ability of the firm to acquire technology. The scope of the system will vary directly with the level of technology and the flow of information.

When a new technique has been adopted, the speed at which other firms adopt may differ widely. This leads to what can be called the rate of diffusion (imitation). The rate of diffusion will be faster, the greater the improvement over existing technology and the lower the cost of the technology in general (Roy and Cross, 1975). Using the definition of Bradley et al. (1995), technology diffusion can be defined as the spread of a new technique from one firm to another ("inter-firm diffusion") (Stoneman and Karshenas, 1993). The two principal types of technology diffusion are "disembodied" diffusion (the transmission of knowledge and technical expertise) and "embodied" diffusion (the introduction into production processes of machinery, equipment and components incorporating new technology) (Papaconstantinou et al., 1995). Research spillovers are the means by which new knowledge or technology developed by one firm become potentially available to others, and the absorptive capacity of the receiving firms will determine the extent to which the technology is incorporated.

The time pattern of adoption and the speed at which it takes place are distinct happenings. The exploration time period when implementing an innovation can provide imitators with a "window of opportunity" to proliferate (Jayanthi, 1998). Empirical studies suggest that the adoption of a new technology follows a bell-shaped, or normal, distribution curve (Norris and Vaizey, 1973). By plotting cumulatively this shows the number of firms that have adopted a new technology in any given year, and the distribution will give an S-shaped curve. (It was Gabriel Tarde who in the *Laws of Imitation* (1903) proposed that adoptions plotted against time assume a normal distribution, or if plotted cumulatively assume the S-shaped curve. (Baker, 1976; Pijpers et al., 2002; UoT, 2004)) An S-shaped distribution, not necessarily derived from a normal distribution, shows the spread of most new technology. There are two general reasons for the occurrence of this distribution.

The diffusion process for firms is a learning process

Firms who are potential users have to become aware of the technology and then attempt to evaluate it. Consequently they may use the technology on a trial basis. The learning process takes place at this stage. Information about the technology has to be disseminated, and as it is adopted by other firms or by the firm on an experimental basis, the information becomes more reliable. The importance of accumulated knowledge and expertise is an important factor determining whether firms are likely to adopt new technology or to act as sources of innovation (Gurisatti et al., 1997). "Bugs" will be overcome, which will in turn reduce the risk of adopting the technology.

An interaction effect occurs for firms

The first formal study of diffusion was the spread of hybrid corn (Grilliches, 1960). The adoption rate in different states in the United States was studied and it was found that there were significant differences between states in the rate of hybrid corn adoption. Logistic growth curves were fitted by Grilliches (1960) to his data and the parameters found from the curves for the different states showed wide variations.

Another formal study of the rate of diffusion was carried out by Mansfield (1961, 1968) who studied the rate of diffusion of 12 innovations in four industries – coal, iron and steel, brewing and rail (Mansfield, 1961, 1968). Although small firms were not included in the analysis, for medium-sized and large firms in most cases, the spread of innovations over time approximated the S-shaped curve. According to Mansfield (1961, 1968) the spread of innovations is best described by a logistic curve. Despite the shape of the curve for technology diffusion appearing S-shaped, there will be differences in the speed at which technology is diffused and the length of the diffusion process. Both within and between industries there will be considerable variations in the rate of the diffusion of technology between firms.

Important factors which appear to affect the rate of diffusion (the speed at which a new technology is accepted) are the characteristics of the firm and the characteristics of the technology itself. Early work on the categories of adopters found that further to adoption following a normal distribution curve the distribution could be used to show the categories of adopters (Rogers, 1962). Firms that are early adopters will tend to be "technically progressive" and will be close to the best that can be achieved in the practice of applying technology (Carter and Williams, 1957).

The speed of diffusion will also be faster the greater the awareness of firms to the advantages of adopting a new technology. The process of communication will be important here as well as the ability of firms to assess the merits of the technological advance. A firm is more likely to adopt a new technology as it diffuses due to being under increasing competitive pressure to do so, through the technology becoming more attractive, and as a result of information about the technology being broadcast from an increasing base (Green and Morphet, 1975). Two basic mechanisms available to firms are technology exchange (technology passed from one firm to another) and technology exploitation (technology transferred to a firm from an external source).

2.3   Technology transfer

Technology transfer is an active process whereby technology is carried across the border of two or more social entities (the external source and the firm), and technology transfer channels are the link between the entities (in which various technology transfer mechanisms are activated) (Autio and Laamanen, 1995). A technology transfer mechanism is defined as any

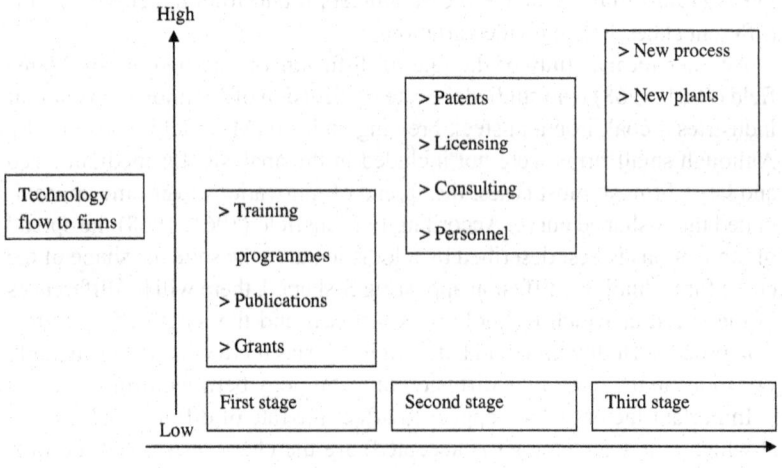

Figure 2.1 Technology transfer mechanisms

Source: Adapted from Chen (1994, p. 451).

specific form of interaction between entities during which technology is transferred (Autio and Laamanen, 1995).

Technology transfer is viewed as an exchange process by Burati and Penco (2001), where a collaborative venture transpires involving a technology donor and recipient working in partnership to adapt and develop technologies (with the aim of dealing with the customisation of technology required to develop specific applications, applying new technology to create value for the recipient taking into account both internal and external factors, and the needs of potential users). Figure 2.1 illustrates technology transfer mechanisms based on the work of Chen (1994) and adapted to new process diffusion. Here the last stage concerns new process and new plants.

2.4 A model of technology diffusion

A model of the diffusion of technology into firms can be described as innovation (supply) from the source of technology (origins) and diffusion (demand) to the firm (destination). A temporal model (Thomas et al., 2001) of technology diffusion measures the speeds of diffusion (or rates of technology transfer) (Bradley et al., 1995).

Although the variables involved in such a model appear to be the most important influences on technology diffusion into firms, there will also be a multiplicity of influences that accelerate or alleviate the rate of diffusion. This spectrum of influences on diffusion rates broadens when considering technology transfer among the various different firms. An extension of the hypothetical example of diffusion is the diffusion of technology into firms. Firms' sociological forces will have an important role to play. The rate of adoption of a new technology will be faster if it is compatible with the previous experience and present normative values of firms. Other influences on the speed of diffusion include the complexity of the new technology and random influences. A model of technology diffusion illustrates that the successful diffusion of a new technology involves considerably more than technical competence. Many complementary factors will be prominent.

2.5 New process diffusion and business partnerships

With new process diffusion and business partnerships it is worth noting the quote, "when there is teamwork and collaboration, wonderful things can be achieved" (Mattie Stepanek, 1990–2004), for businesses pressures include global competition, short product life cycles and technological change (which have transformed their competitive environment) (Ali, 1994; Bettis and Hitt, 1995). Within this context, Autio and Laamanen (1995, p. 647) talk about the ability to recognise technical problems, the ability to develop new concepts and tangible solutions to technical problems, the concepts and tangibles developed to solve technical problems, and the ability to exploit the concepts and tangibles in an effective way.

Various types of inter-organisational relationships undertaken in practice are reported in the literature, and these include interlocking directorates, trade associations, alliances, consortia, networks and joint ventures. These vary according to partnership linkages (Barringer and Harrison, 2000).

Formalisation of an agreement can exist for personal formal relationships and third parties, while in remaining groupings formalised relations are evident (Bonaccorsi and Piccaluga, 1994). The issue of formalisation is considered to be significant since formalisation and monitoring of inter-organisational relationships can cause disagreement and loss of trust among partners through them attempting to retain independence for their organisations in a situation where interdependence is increasing (Ring and van de Ven, 1994).

From the literature on inter-organisational relationships between 1960 and 1990, six critical contingencies have been posited by Oliver (1990) across linkages, settings and organisations; these are necessity, asymmetry, reciprocity, efficiency, stability and legitimacy (Oliver, 1990). According to Oliver (1990), two delimiting assumptions are behind the determinants: deliberate decisions are assumed to be made to form an inter-organisational relationship by organisations, and an organisational perspective involving a top management approach is assumed (the determinants can also explain lower reasons) (Oliver, 1990). The six contingencies show strong correlation with alliance strategy motives (Eisenhardt and Schoonhoven, 1996). The exploitation of intellectual property rights, licensing of patents and the commercialisation of research will be important. There will be motivations for businesses to have inter-organisational relationships from a standpoint of efficiency (Ankrah, 2007).

With regard to the business inter-organisational relationship, a number of typologies have been developed to express the diversity of relationships that may be employed in the collaborative process. Freeman (1991) distinguishes between the following: joint ventures and research corporations; joint R&D agreements; technology exchange agreements; direct investment motivated by technology factors; licensing and second-sourcing agreements; sub-contracting, production-sharing and supplier networks; government-sponsored joint research programmes; computerised databanks for technical and scientific interchange; and informal or personal networks. Concerning the case study of the management of the early float glass start-ups, patents and licensing played an important role.

References

Ali, A. (1994) Pioneering versus incremental innovation: Review and research propositions, *Journal of Product Innovation Management*, 11, 46–61.

Ankrah, S. N. (2007) *University-Industry Interorganisational Relationships for Technology/Knowledge Transfer: A Systematic Literature Review*, Leeds University Business School Working Paper Series, 1(4), June.

Autio, E. and Laamanen, T. (1995) Measurement and evaluation of technology transfer: Review of technology transfer mechanisms and indicators, *International Journal of Technology Management*, 10(7/8), 643–664.

Baker, M. J. (1976) Chapter 7, Diffusion theory and marketing, in *Marketing Theory and Practice*, London: Macmillan, pp. 119–131.

Barringer, B. R. and Harrison, J. S. (2000) Walking a tightrope: Creating value through interorganizational relationships, *Journal of Management*, 26(3), 367–403.

Bettis, R. and Hitt, M. (1995) The new competitive landscape, *Strategic Management Journal*, 16, 7–19.

Bonaccorsi, A. and Piccaluga, A. (1994) A theoretical framework for the evaluation of university-industry relationships, *R&D Management*, 24(3), 229–247.

Bradley, A., McErlean, S. and Kirke, A. (1995) Technology transfer in the Northern Ireland food processing sector, *British Food Journal*, 97(10), 32–35.

Brooksbank, D., Morse, L., Thomas, B. and Miller, C. (2001) Technology diffusion, *Entrepreneur Wales*, *Western Mail*, 8–9.

Burati, N. and Penco, L. (2001) Assisted technology transfer to SMEs: Lessons from an exemplary case, *Technovation*, 21(1), 35–43.

Carter, C. and Williams, B. (1957) *Industry and Technical Progress*, London: Oxford University Press.

Chegg (2018) Definition of Manufacturing Processes, www.chegg.com/homework-help/definitions/manufacturing-processes-5.

Chen, E. Y. (1994) The evolution of university-industry technology transfer in Hong Kong, *Technovation*, 14(7), 449–459.

Eisenhardt, K. M. and Schoonhoven, C. B. (1996) Resource based view of strategic alliance formation: Strategic and social effects in entrepreneurial firms, *Organisation Science*, 7(2), 136–150.

Freeman, C. (1991) Networks of innovators: A synthesis of research issues, *Research Policy*, 20(5), 499–514.

Green, K. and Morphet, C. (1975) Section 7, The diffusion of innovations, in *Research and Technology as Economic Activities*, York: Science in a Social Context (SISCON), pp. 45–47.

Grilliches, Z. (1960) Hybrid corn and the economics of innovation, *Science*, July 29, 275–280.

Gurisatti, P., Soli, V. and Tattara, G. (1997) Patterns of diffusion of new technologies in small metal-working firms: The case of an Italian region, *Industrial and Corporate Change*, 6(2), March, 275–312.

InnoviSCOP (2018) Process innovation Definition, www.innoviscop.com/en/definitions/process-innovation.

Islam, T. and Meade, N. (1997) The diffusion of successive generations of a technology: A more general model, *Technological Forecasting and Social Change*, 56(1), 49–60.

Jayanthi, S. (1998) *Modelling the Innovation Implementation Process in the Context of High-Technology Manufacturing: An Innovation Diffusion Perspective*, Cambridge: ESRC Centre for Business Research.

Mansfield, E. (1961) Technical change and the rate of imitation, *Econometrica*, October, 741–766.

Mansfield, E. (1968) Chapter 4, Innovation and the diffusion of new techniques, in *The Economics of Technological Change*, New York: Norton, pp. 99–133.

Norris, K. and Vaizey, J. (1973) Chapter 7, The diffusion of innovations, in *The Economics of Research and Technology*, London: George Allen & Unwin, pp. 86–103.

Oliver, C. (1990) Determinants of interorganisational relationships: Integration and future directions, *Academy of Management Review*, 15(2), 241–265.

Papaconstantinou, G., Sakurai, N. and Wyckoff, A. W. (1995) *Technology Diffusion, Productivity and Competitiveness: An Empirical Analysis for 10 Countries, Part 1: Technology Diffusion Patterns*, Brussels: European Innovation Monitoring System (EIMS).

Pijpers, R. E., van Montfort, K. and Heemstra, F. J. (2002) Acceptable van ICT: Theorie en een veldonderzoek onder top managers, *Bedrijfskunde*, 74(4).

Rialp-Criado, A., Rialp-Criado, J. and Knight, G. A. (2002) *The Phenomenon of International New Ventures, Global Start-Ups, and Born-Globals: What Do We Know after a Decade (1993–2002) of Exhaustive Scientific Inquiry?* Working Paper, Department d'Economia de l'Empresa, Universitat Autònoma de Barcelon, Barcelona.

Ring, P. S. and van de Ven, A. H. (1994) Developmental processes of cooperative interorganisational relationships, *Academy of Management Review*, 19(1), 90–110.

Rogers, E. (1962) *Diffusion of Innovations*, New York: Collier-Macmillan.

Roy, R. and Cross, N. (1975) Section 3.1.3, Diffusion, in *Technology and Society*, T262 2–3, Milton Keynes: Open University Press, pp. 36–38.

Schon, D. A. (1971) Chapter 4, Diffusion of innovation, in *Beyond the Stable State*, London: Temple Smith, pp. 80–115.

Stoneman, P. and Karshenas, M. (1993) The diffusion of new technology: Extensions to theory and evidence, in *New Technologies and the Firm: Innovation and Competition* (ed. P. Swann), London: Routledge, pp. 177–200.

Thomas, B., Packham, G. and Miller, C. (2001) A temporal model of technology diffusion into small firms in Wales, *Industry and Higher Education*, August.

University of Twente (UoT) (2004) Diffusion of Innovations Theory, *Theorieenoverzicht TCW*, www.utwente.nl/cw/theorieeboverzicht.

Wakkee, I., van der Sijde, P. and Kirwan, P. (2003) *An Empirical Exploration of the Global Startup Concept in an Entrepreneurship Context*, GS Leuven, Working Paper.

3 Case study – management of the early float glass start-ups

3.1 Background

The company that successfully developed the float glass process was Pilkington Brothers Limited, a long-established British company (Jewkes et al., 1969) founded in 1826 by a group of local entrepreneurs in St Helens, which started manufacturing glass at the Crown Glass Company, with the company being renamed Pilkington in 1849 (Davis, 2011). According to Jevons (1973), the float glass process (Pilkington, 1963) appears to be a technological discovery since it cannot be explained according to the application of a specific discovery by curiosity-oriented scientists or an academic. The original concept for the manufacture of flat glass using molten tin was patented in the United States by Heal (1902).

As related by Utterback (1996), Pilkington had made plate glass by the Ford/Pilkington method for many decades and had introduced process innovation on an incremental basis. An example of this was the development of continuous grinding and polishing. Pilkington therefore had historically undertaken considerable innovation activities with the continuous grinding and polishing process in the 1920s, which had been licensed to plate glass manufacturers, and a twin machine which allowed plate glass to be polished on both sides at the same time in 1935 (Davis, 2011). Pilkington brought its first twin grinder into use at its Doncaster works, giving the company global leadership at a technological level for large-scale plate glass manufacture (Utterback, 1996). This showed that Pilkington had the reputation and experience for creating and developing new glass manufacturing systems (Davis, 2011).

Glass manufacture was a stable industry with a low rate of innovation involving Pilkington the British market leader, there was no pressing competitive threat causing them to innovate, and the stimulus to innovate came from within the company (Twiss, 1979). In January 1947 a body called the Manufacturing Conference was set up by James Meikle, Pilkington senior production director, to meet monthly, and at the March meeting Meikle asked whether efforts should be made to develop a new method of plate glass production without costly grinding and polishing (Barker, 1994). The team of engineers formed first investigated the use of vibrating platens or plates instead of grinding and polishing, which was an invention of the Owens-Illinois Glass Company's (Toledo, Ohio) director of research (Barker, 1994). Although this did not lead to improvements, the part of the process that could be changed was identified by Pilkington which would result in considerable returns (Barker, 1994).

By the mid-1950s there was considerable competition developing in the market through experimentation with new production methods and techniques by Asahi Glass in Japan, Ford Motor Company in the USA, Pittsburgh Plate Glass Company also in the USA, and St Gobain in Belgium (Davis, 2011). It was therefore the case that whichever company developed a process to produce cheap good quality glass on a continuous basis would attain market leadership (Davis, 2011). This resulted in the summer of 1949 with Pilkington establishing two development teams; Alastair Pilkington, whose father was distantly related to the Pilkington family and had recommended his son to the company, led one of these teams (Davis, 2011). As a member of the family, Alastair Pilkington, born in 1920, was a graduate of the University of Cambridge in Mechanical Science, joined Pilkington in 1947 and later was appointed to be in charge of plate glass production in 1955 as director (Jewkes et al., 1969).

The company was therefore considering replacing grinding and polishing when Alastair Pilkington was appointed to Doncaster in 1949 as production manager when the experiments with platens commenced in St Helens. There was the need to discover a way of handling the soft ribbon of glass from the platens, which led him to floating a glass ribbon over a molten metal surface (Barker, 1994).

At this time, in the mid-twentieth century, Pilkington transformed itself from a national firm into a global leader (Grundy, 1990) through innovations in glass manufacture (Davis, 2011). Before this time glass manufacture was undertaken using one of two techniques – either plate glass, which produced good quality glass but was expensive and slow due to machine and labour costs, and sheet glass, which although inexpensive and fairly fast to produce was not of the same quality (Davis, 2011). The goal was therefore to develop a technique to produce glass with the quality of plate glass but at the price of sheet glass, and this was especially the case with the motor industry, which wanted better quality mass-produced glass for windscreens at a competitive price (Davis, 2011).

3.2   Process development

The research team led by Alastair Pilkington, based at the Cowley Hill plant of the company in St Helens, commenced by investigating a glass making system in development in the US called the Bowes process, which smoothed hot glass when it came from the rollers still soft (Davis, 2011). In response to the difficulty of moving glass along the production line when still cooling and liable to mark, one of the engineers on Pilkington's team, Ken Bicker-staff, suggested using liquid metal for transporting the glass (Davis, 2011). As a result, experiments commenced at the Doncaster plant using molten tin, although in theory molten glass and molten tin did not mix and when together had an adverse reaction (Davis, 2011). For the experiments on the use of tin as a carrier at a temperature of around 600 degrees Celsius, under-taken by Ken Bickerstaff, at the time there was no question of forming the ribbon over molten tin at a higher temperature of some 1,000 degrees Cel-sius (Barker, 1994).

Returning to St Helens, Alastair Pilkington foresaw the possibility of floating the ribbon along the bath of molten tin at a temperature gradient of 1,000 degrees Celsius down to 600 degrees Celsius, where rollers could take it forward (Barker, 1994). This suggestion was made by Alastair Pilk-ington at a regular meeting of the Manufacturing Conference, and as work on the platens was not making progress, it was agreed that the engineering team should explore the LAB project, named using the initials of Alastair Pilkington (Barker, 1994).

Building on Ken Bickerstaff's idea, two breakthroughs occurred in 1952: (1) Alastair Pilkington realised that molten tin would smooth glass direct from the rollers, eliminating the expense and cost of the Bowes process, and (2) the idea of "freefall", where pouring liquid onto a surface that is flat naturally forms a pool, meaning that molten glass poured onto molten tin flows in a natural state until a quarter of an inch thick, which is the required thickness for the manufacturing process (Davis, 2011). Alastair Pilkington had therefore inspired a programme of R&D in 1952 that would lead to a revolutionary method of glass manufacture (Utterback, 1996). The float glass process enabled molten glass from a furnace to form on a surface of molten tin that was perfectly flat (since tin was more dense it would support the glass and be nonreactive) (Utterback, 1996).

Early samples of glass were made on a 12-inch wide pilot plant in 1953, and there were a number of challenging development stages, including in 1954 when it appeared that it would not be possible for the surface between glass and metal to attain the necessary standard (Jewkes et al., 1969). Con-struction of the pilot plant had cost £25,000 and by 1954 the early costs were over £100,000; following this, costs began to soar (Jewkes et al., 1969). The theory upon which float is based, therefore, was only the beginning, since it took seven years of investment and research prior to Pilkington being able to "go public" (the estimated cost for the company was £7 million, equivalent to £80 million today, for the development of the process which involved many difficulties) (Davis, 2011). The spring of 1955 saw the Pilkington board authorise a full-scale production unit (Barker, 1994). The Cowley Hill (CH1) float tank was lit in 1957 on May 6, with the backing of the company (Jewkes et al., 1969); there had already been costs of £676,000 and it did not make a saleable square foot of glass for 14 to 15 months (Jewkes et al., 1969; Barker, 1994). Costs ran at about £100,000 per month (£1.14 million in today's money), and the Pilkington board met frequently in order to discuss continuing funding process development (Davis, 2011).

Fortunately they agreed to continue funding, and in July 1958 a square foot of good glass was produced (Jewkes et al., 1969), with Sir Harry Pilkington (later Lord Pilkington) on January 20, 1959, announcing float glass to the world, although the quantity of the glass produced was "quite considerable" but "not yet enough to be offered freely" (Barker, 1994). By the time the first saleable glass had been made, development costs were at around £3 million, reaching £7 million in development costs when the float glass process reached the stage to replace all plate glass (Jewkes et al., 1969).

It was in September 1959 that the new float glass process at the Cowley Hill plant was able to consistently produce high quality glass (Davis, 2011). Although anxieties did not end here, as noted by Pilkington (1966):

> As the set-up on the float bath was rather old and tattered. . . .We renewed the worn out parts and expected to settle down to a long successful run. To our amazement we then made continuous cullet (bad glass) again and were back struggling and feeling more frustrated than ever before. It took us nearly three months of investigation to discover that when we first made good glass it was partly due to a fluke and that a vital part of our success had been due to a broken part of the plant. As soon as we understood the problem we made a set-up which reproduced similar conditions to those with the broken part. This made saleable glass immediately and we have never been in doubt again about the ability of the process to make consistently good glass.
>
> (Pilkington, 1966)

The "float" glass team had a second piece of good luck, since it became apparent that whatever thickness rolled the final sheet of glass would be a quarter of an inch thick, and this was the thickness of approximately half of company sales (Jewkes et al., 1969). Consequently, Pilkington was able to sell glass in large quantities profitably while developing techniques to produce thinner and thicker glass (Jewkes et al., 1969). One of the first sheets of float glass to be successfully produced by the process is shown in Figure 3.1 with the key people involved in the development team, and Figure 3.2 is a picture of one of the first two sheets successfully produced.

Figure 3.1 The original "float" glass team, 1959

Source: Barker (1994) and James Edward Celfyn Thomas' archive.

Note: Alastair Pilkington (far left) demonstrating the distortion-free quality of a small piece of float glass to (on his left) Ernie Litherland, production manager, Cowley Hill; George Dickenson, development manager; James Edward Celfyn Thomas, tanks manager; Jack Topping, special examiner; and Richard Barradell-Smith (ex–Rolls Royce), leader of the float development team.

Figure 3.2 Picture of one of the first two sheets of float glass successfully produced

Source: Photograph taken of sheet of float glass from James Edward Celfyn Thomas' archive.

In the photograph in Figure 3.2, the label on the outside of the box containing the sheet of float glass reads:

Sk 383-K23 79542

To Mr. J.E.C. Thomas
Tanks Department
COWLEY HILL WORKS

From PILKINGTON BROTHERS LIMITED, Glass Manufacturers,
ST. HELENS, LANCS.

On the actual sheets of float glass the letter to the bottom left of the picture reads:

FROM   PILKINGTON BROTHERS LIMITED   TELEPHONE:
SIR HARRY   GLASS MANUFACTURERS.   ST.HELENS 4001
PILKINGTON   ST HELENS. LANCS

February, 1959

Dear Sir,

We have had a number of pieces of Float Glass engraved for individuals who have been specially concerned in production, who occupy a senior position in the Firm, or are otherwise specially interested, and I have pleasure in enclosing one for you personally.

Yours sincerely,
Harry Pilkington

At the top of the sheet of float glass it is stated:

NO.1 TANK FLOAT GLASS, COWLEY HILL WORKS, NOVEMBER 1959

Pilkington had taken a considerable number of years and many millions of pounds in development work when they announced the process in1959 (Jevons, 1973). Scientists had become involved in the process to resolve technological problems and to obtain a greater understanding of the process, following the development of the process empirically to a stage for the production of satisfactory glass (one example was the cause of a surface bloom was found to be a layer of tin dissolved in the glass) (Jevons, 1973). This work used analytical techniques and fundamental concepts which could have been from curiosity-oriented research (Jevons, 1973).

3.3 Process diffusion through patenting, licensing and the early start-ups

It was in July 1962 that the first license was issued for the float glass process. In the financial year ending March 1963 the second float line (CH4) commenced in St Helens, and the new product of float glass brought in a steady profit to Pilkington (Barker, 1994). By 1967 Pilkington discontinued plate glass production, and in 1970 there were 28 float glass plants operating abroad (Barker, 1994). Figures 3.3 and 3.4 provide a comparison of the stages of the plate glass process and float glass process.

The plate glass process involved casting a plate of glass, grinding it flat and polishing to make it transparent, which were expensive and lengthy processes, and resulted in removing the natural hard surface with loss of 20% of good glass ground away to obtain necessary flatness (Jewkes et al., 1969).

With the float glass process, a continuous ribbon of glass from the melting furnace floats on the surface of a bath of molten tin with a flat surface to produce glass that is flat, and the ribbon cooled on the molten tin and passed

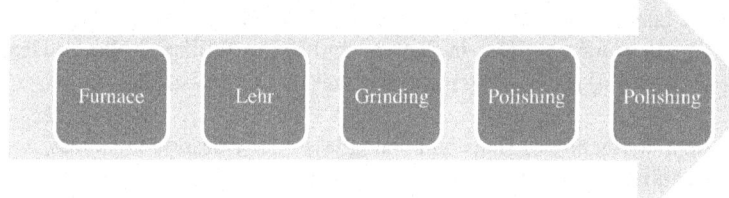

Figure 3.3 Polished plate glass
Source: Grundy (1990).

Figure 3.4 Float glass
Source: Grundy (1990).

through an annealing oven to produce uniform thickness with polished surfaces without grinding and polishing (Jewkes et al., 1969). There were considerable advantages to the process, since costs in forming a ribbon from a stream of molten glass were reduced, there was no loss of glass in grinding, a continuous process replaced a non-continuous process and variations in glass width were cheaper and quicker to make (Jewkes et al., 1969). Utterback (1996) further noted that "besides improved quality the new process produced a remarkable set of production efficiencies" (p. 112). In fact, the float glass plant brought together all aspects of automation within a single continuous process, and transformed plate glass making from being labour intensive into an automated industry with high efficiency (Utterback, 1996).

R&D for the float glass process was not straightforward but there were considerable potential rewards, and Pilkington had been ensured that key process inventions were patented in many countries before the breakthrough announcement (Davis, 2011). With regard to patent protection, the first British patent was applied for by Pilkington in December 1952, specifications submitted in 1954, and in July 1957 the patent was sealed (Pilkington and Bickerstaff, 1957). Following this, the US patent was issued in November 1959 (Pilkington, 1959), having been applied for in December 1954, but the US Patent Office had drawn attention to two earlier US patents when examining the later application (Jewkes et al., 1969). The US Patent Office had issued a patent for a process in September 1902 to William Ephraim Heal (Marion, Indiana) (Heal, 1902), and this was described as follows:

> The manufacture of sheet and plate glass of any desired thickness, and in continuous sheets by a new and improved method of flowing the molten glass from the melting tank into an adjacent receptacle, containing melted material of a greater specific gravity than glass (preferably of tin or alloys and copper) and causing the molten glass to float upon and spread into, a continuous sheet, and then drawing the sheet of glass therefrom, and causing it to pass into their lehr for annealing and by one continuous operation . . . to simplify, facilitate and cheapen the manufacture of sheet and plate glass to improve its quality.
>
> (Heal, 1902)

A patent for the same process had been applied for, and granted in January 1903, in the United Kingdom (Heal, 1903). The other US patent had been granted to H. K. Hitchcock (Walton, Pennsylvania) in May 1905, and was described as follows:

> The plastic glass. . . . forced through a slot in the side wall of the furnace. . . . enters a chamber in which it is hardened and annealed.

The support for the sheet or plate of glass. . . . consists of a practically continuous liquid bed, formed of molten metal. . . . so that the sheet or plate will float on the surface of the molten metal.

(Hitchcock, 1905)

Since these two inventors emphasised floating molten glass on a bath of molten tin, this led to scrutiny of the Pilkington application by the US patent examiners, and to reject some of its claims (patent examiners in the United Kingdom did not undertake searches more than 50 years, whereas in the United States this was not the case) (Jewkes et al., 1969). Heal (1902, 1903) and Hitchcock (1905) had not evolved an operative production process, had not undertaken experimental work on a large scale, and through their ideas had not produced satisfactory glass (Jewkes et al., 1969). Importantly, the success of the float glass process was based on a continuous flow of glass from a furnace, which had not been achieved until the 1920s when Pilkingtons did this (Jewkes et al., 1969). The advantages of the process were realised globally (Grundy, 1990) since by the close of 1967 Pilkington's licensed patents with knowledge for formation of a glass ribbon on a horizontal plane without rollers to manufacturers in many countries including the United States, Belgium, France, Italy and Germany (Earle, 1967).

Not only were inventions and patents important since Pilkington engineers' considerable knowledge about the float glass process and detailed information were required for perfect glass production (recorded in manuals and engineering drawings with personnel bound by agreements on confidentiality) (Davis, 2011). Pilkington's control over the idea of the float glass process was not only legal but also dependent on access to their files, meaning that without these the process was unworkable (Davis, 2011).

The belief by Pilkington that the float glass process would be the dominant method of producing glass was realised, and as a result all the companies' rivals bought licenses from them to use the process, which included first the Pittsburgh Plate Glass (PPG) Company, quickly followed by Libbey Owens Ford (LOF) and the Ford Motor Company (Davis, 2011). Many years after original patents expired, the company was able to obtain license fees through enforcing protection of its technological advancements, with the process eventually licensed in 30 countries to 42 manufacturers (Davis, 2011). By this time the company earned around $4–5 bn (approximately £3bn or £4bn today) from licensing the process, and the company moved from being the fifth largest glass manufacturer in the world in 1959 to becoming the global leader – illustrating the rewards of R&D, innovation and high-value manufacturing (Davis, 2011). An important linkage in the diffusion of the process through patenting and licensing was the progress made from the first workable process in 1959 and the early float glass start-ups, which enabled the process to be diffused by Pilkington to the leading glass manufacturers. In order for this to be successful, a number of the leading glass technologists involved in the development of the process enabled successful diffusion through the management of the early float glass start-ups. These included Alastair Pilkington at board level, James Edward Celfyn Thomas at management level for the early float glass start-ups and Tom Grundy at glassmaker level. In terms of the management of the early float glass start-ups, this case study is focussed through the lens of Celfyn Thomas, who was in charge of these early start-ups. As related in *Triplex* magazine, when Celfyn Thomas was appointed works manager at Eccleston on November 15, 1966:

> Mr. Thomas was previously float glass manufacturing manager at Pilkington Brothers' Cowley Hill Works, where he started up the world's first float glass plant in 1957.
>
> Having graduated at Cambridge with an honours degree in engineering, Mr. Thomas joined Pilkington Brothers in 1951 as a technical assistant in Doncaster. In 1954 he was transferred to St. Helens, where he became deputy manager of plate glass manufacturing for a time.
>
> Mr. Thomas has been responsible for establishing float glass plants in Maryland, for the Pittsburgh Plate Glass Company, and in California, France, Belgium and Japan. He was senior production manager with a team which designed the St. Gobain float plants at Pisa, Italy, and Porz, near Cologne; also a Ford plant in Tennessee. He was presented to the Queen and Prince Philip when they visited Pilkingtons to see the float process in 1961 and to Princess Margaret and Lord Snowdon in 1963.
>
> (Triplex, 1966)

Celfyn Thomas was a co-inventor on two of the float glass patents. These included the UK patent titled "Improvements in or Relating to the Manufacture of Flat Glass" (Dickinson et al., 1968). The original patent specification with attached drawings was number 1,112,071 with a date of filing the complete specification of February 3, 1966, an application date of February 25, 1965, and the complete specification published May 1, 1968. The following is stated at the start of the complete specification:

> We, PILKINGTON BROTHERS LIMITED, a British Company of 277/283 Martins Building, Water Street, Liverpool 2, Lancashire, England, GEORGE ALFRED DICKINSON, a British Subject of Bryn Mair, Prescot Road, St. Helens, Lancashire, England, JOHN HENRY MORGAN, A British Subject of 61 Thornton Road, Liverpool 16, Lancashire, England, JAMES EDWARD CELFYN THOMAS, a British

Subject of 4 Rosebury Road, St. Helens, Lancashire, England, and BRIAN WILLIAM OXLEY, a British Subject of Grassy Corner, Davy Close, Eccleston, St. Helens, Lancashire, England, do hereby declare the invention, for which we pray a patent may be granted to us, and the method by which it is to be performed, to be particularly described in and by the following statement:–

This invention relates to methods of making flat glass on a molten metal bath by flowing glass in a molten state on to the bath and advancing the glass along the bath until it is cooled sufficiently to be taken unharmed from the bath surface. Desirably the molten metal bath is a bath of molten tin or molten tin alloy having a specific gravity greater than glass and in which tin predominates and preferably the bath is so constituted as to have all the characteristics described in British Patent No. 769,692.

The other was the US patent entitled "Apparatus and Method for Manufacture of Float Glass with Restricted Lateral Spread" (Dickinson et al., 1969), stating George Alfred Dickinson, St Helens; John Henry Morgan, Liverpool; and James Edward Celfyn Thomas and Brian William Oxley, St Helens, England, assignors to Pilkington Brothers Limited, Liverpool, England, a corporation of Great Britain, US Patent Office, number 3,433,612, patented March 18, 1969. The patent was filed February 8, 1966 with Serial No. 525,976. The abstract of the disclosure states:

Molten glass is delivered to a bath of motel metal, is initially laterally contained on the bath and is then permitted restricted lateral spread controlled by surfaces of a material wetted by the molten glass as a ribbon of float glass is developed.

Figure 3.5 is a picture of Celfyn Thomas being presented to the Queen by Sir Harry Pilkington, and Figure 3.6 is a picture of what is believed to be the gloves used by Sir Harry Pilkington to handle the first sheets of float glass.

3.4 Managing the early start-ups

Following the first foreign license being issued in 1962 to the Pittsburgh Plate Glass Company (Henry and Walker, 1991) the early float glass start-ups took place. Table 3.1 lists some of the first licenses (Grundy, 1990) leading to the early float glass start-ups which were managed by Celfyn Thomas.

With regard to the early float glass start-ups in Japan, Figure 3.7 shows Alastair Pilkington and Celfyn Thomas at a meeting, Figure 3.8 Celfyn

Figure 3.5 Picture of Celfyn Thomas being presented to the Queen

Source: Photograph from James Edward Celfyn Thomas' archive (see Appendix 2).

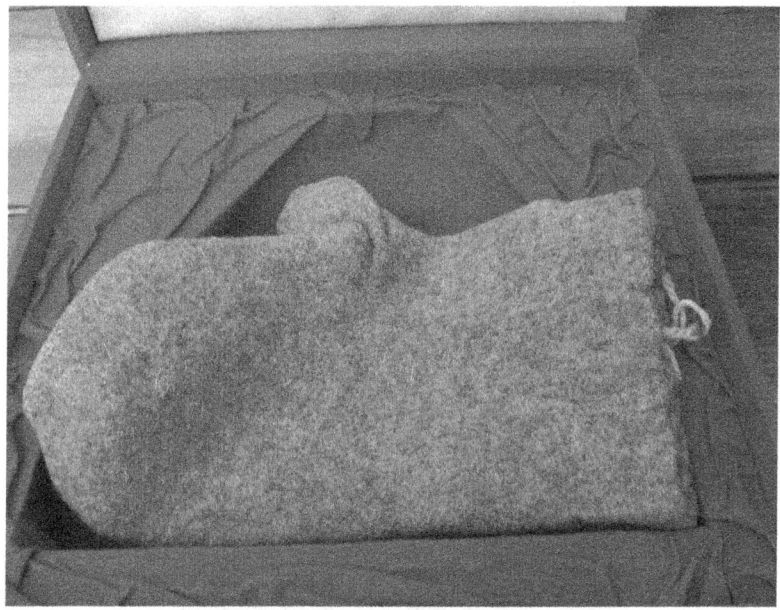

Figure 3.6 Gloves probably worn by Sir Harry Pilkington to handle the first sheets of float glass

Source: Photograph taken of gloves from James Edward Celfyn Thomas' archive.

Table 3.1 Licenses leading to the early float glass start-ups

Date	License
July 27, 1962	Pittsburgh Plate Glass, USA
December 14, 1962	Glaverbell, S.A.
April 19, 1963	Libbey Owens Ford, USA
January 6, 1964	Compagnie de St Gobain, France
March 6, 1965	Asahi Glass Co. Ltd., Japan
March 6, 1964	Nippon, Japan
July 29, 1964	Ford Motor Company, USA

Source: Grundy (1990).

Thomas inspecting a sheet of float glass with a glass worker, and Figure 3.9 Alastair Pilkington and Celfyn Thomas with a Japanese float glass team.

For the management of the early float glass start-ups there were a number of important factors that were evident with regard to successfully starting

Figure 3.7 Alastair Pilkington and Celfyn Thomas at a meeting
Source: James Edward Celfyn Thomas' archive.

Figure 3.8 Celfyn Thomas inspecting a sheet of float glass with a glass worker
Source: James Edward Celfyn Thomas' archive.

Figure 3.9 Alastair Pilkington and Celfyn Thomas with a Japanese float glass team
Source: James Edward Celfyn Thomas' archive.

up and enabling the process to operate efficiently. These included a deep
knowledge of the process from many years of involvement in its development, a high level of expertise in overcoming production line problems
especially with regard to tanks management, leadership of glass production
workers and the ability to work with glass workers and technologists from
other countries globally.

3.5 Overview

The float glass process is now adopted by all major manufacturers of
glass, being a new concept when it was developed by Pilkington Brothers Limited, where molten glass fed continuously onto the surface of a
bath of molten tin on which it floats produced a constant thickness and
smooth surface for the resultant glass, removing the need for grinding
(Twiss, 1979). Even though the concept of floating glass on tin is simple,
the practical development of a mass production process had major time
and cost implications (Twiss, 1979). According to Barker (1994) there
were two important conditions for the successful development of the float
glass process: first, world glass demand for both building and the motor

industry was growing rapidly, and in these circumstances the decision for float investment in 1955 was timely; and second, the presence on the board of Alastair Pilkington where he could make the case for continued development was vital to the success of the process. Float glass, developed by Pilkington Brothers Limited, involved an essentially new process, first announced in 1959 and following further development licensed on a worldwide basis (Barker, 1994). By largely reducing costs it quickly replaced the manufacture of plate glass, and following further development it replaced sheet glass (Barker, 1994).

Alastair Pilkington, who was the project champion, was able to pass his enthusiasm to those working alongside him and this is evident in his account (Pilkington, 1969) where he says "one of the most exciting things in the history of the flat glass industry" is where "we were all tremendously excited and enthusiastic" and "to be a good development man you must be a born optimist" (Twiss, 1979, p. 20). As well as being a project champion and having an idea that was creative, there was the need for Pilkington's senior management to sanction the immense costs which during development were continuing to escalate (Twiss, 1979). Here "the board were only interested in a cold blooded, objective analysis of the project and progress" (Twiss, 1979, pp. 20–21). These were satisfied, and "in 1954 the Pilkington board decided to give the project the highest possible priority, so that success or failures would be decided as early as possible" (Twiss, 1979, p. 21). By accepting the risks, Alastair Pilkington recounts, "however, at the time when the board's decision was taken it seemed quite likely that glass coming into contact with metal would always be spoiled too much to make it a stable product" (Twiss, 1979, p. 21). At one point in time it appeared the project had serious technical difficulties, but the board stood by and as Alastair Pilkington notes:

> on our production plant we made unusable glass for one year and two months. I had to report regularly to the board, and every month put in a requisition to justify another month's expenditure of £100,000. It was a tremendous credit to the board that they gave unwavering support throughout.
>
> (Twiss, 1979, p. 21)

Alastair Pilkington further comments with regard to the decision for the construction of the first gloat glass production plant: "we thought we were very much more knowledgeable than we turned out to be", and "in retrospect we were woefully unaware of the magnitude of the problems we were going to face when we reached a mass production scale" (Twiss,

1979, p. 21). With regard to this, Roberts (1969, 1970) concludes successful innovators rather than research are development oriented since research leads to further process or product development, as in the float glass process. Difficulties with the float glass process arose later in its development; if the construction of the production plant had been delayed to have greater technological understanding, although costs would have been reduced, most probably there would have been delay in commercial launch (Twiss, 1979).

Promising results had been realised through three small pilot plants, and following a profitable period for the company a full-scale production unit was started up in May 1957 by Pilkington (Barker, 1994). There were considerable theoretical problems and scientists within the company had predicted failure, however the engineers continued to work on the process, and as a member of the executive committee, Alastair Pilkington was able to make the case to other directors on the board (Barker, 1994). Even though the plant was unable to produce saleable glass for 15 months, there was support from Sir Harry Pilkington, who was chairman. On January 20, 1959, float glass was announced to the world, with further development prior to July 1962 with the issuing of the first license (Barker, 1994). The project was an exceptional success, with reductions in plant size of more than a third and costs of production of a quarter, with a cost of seven years of development work and some £4 million for the manufacture of saleable glass (Twiss, 1979).

Comparing Pilkington to the major European competitor St Gobain, which had supplied more than half of Europe's demand for plate glass at the time, in 1962, when Pittsburgh Plate Glass (PPG) Company was the first licensee of float glass, a new US plate glass factory at Kingsport, Tennessee, was opened by St Gobain. It was reported by Smith (1965) that "at the opening ceremonies, PPG's President David G. Hill, letting his eyes rove over the glittering $40 million installation was overheard to remark, 'You are now looking at the most modern obsolete process in the world!'" This led St Gobain to become another Pilkington float glass licensee (Twiss, 1979). By 1966 Pilkington was responsible for 90% of flat glass and 79% of safety glass in the British home market (Jewkes et al., 1969).

Policy on float licensing was to license existing producers and not to allow them to be at a disadvantage through licensing new producers, and Pilkington assisted licensees by allowing them use of improvements as long as these were also available to Pilkington (Barker, 1994). The central thesis of this study, therefore, is that it is not only important to have invention of a process and development through its innovation but also for it to be successfully diffused, as was the case with the early float glass start-ups.

References

Barker, T. C. (1994) *Pilkington an Age of Glass: The Illustrated History*, London: Boxtree.

Davis, E. (2011) *Made in Britain*, London: Little, Brown.

Dickinson, G. A., Morgan, J. H., Thomas, J.E.C. and Oxley, B. W. (1968) *Patent Specification*, London No. 1,112,071.

Dickinson, G. A., Morgan, J. H., Thomas, J.E.C. and Oxley, B. W. (1969) *Patent Specification*, United States No. 3,433,612.

Earle, K.J.B. (1967) The development of the float glass process and the future of the glass industry, *Chemistry and Industry*, July 15.

Grundy, T. (1990) *The Global Miracle of Float Glass: A Tribute to St Helens and Its Glass Workers*, St Helens: Chalen Press.

Heal, W. E. (1902) *US Patent 710*, p. 357.

Heal, W. E. (1903) *Patent Specification*, London No. 19829.

Henry, J. and Walker, D. (1991) *Managing Innovation*, London: Sage.

Hitchcock, H. K. (1905) *US Patent*, US Patent Office.

Jevons, F. R. (1973) *Science Observed: Science as a Social and Intellectual Activity*, London: George Allen & Unwin.

Jewkes, J., Sawers, D. and Stillerman, R. (1969) *The Sources of Invention*, 2nd edn, London: Macmillan.

Pilkington, L.A.B. (1959) *United States Patent Office File Relating to United States Patent*, No. 2,911,759.

Pilkington, L.A.B. (1963) The development of float glass, *Glass Industry*, 44(2), February, 80.

Pilkington, L.A.B. (1966) *Float glass*, *Advance*, November.

Pilkington, L.A.B. (1969) The float glass process, *Proceedings of the Royal Society London*, A314, 1–25.

Pilkington, L.A.B. and Bickerstaff, K. (1957) *Patent Specification*, London No. 769,692.

Roberts, E. B. (1969) Entrepreneurship and technology, in W. H. Gruber and D. G. Marquis (eds.), *Factors in the Transfer of Technology*, Cambridge, MA: MIT Press.

Roberts, E. B. (1970) How to succeed in a new technology enterprise, *Technology Review*, 72(2), December.

Smith, R. A. (1965) At St Gobain, the first 300 years were the easiest, *Fortune*, 72(4), October.

Triplex (1966) Mr. J.E.C. Thomas, *Triplex Magazine*, December 12, p. 20.

Twiss, B. C. (1979) *Managing Technological Innovation*, London: Longman.

Utterback, J. M. (1996) *Mastering the dynamics of innovation*, Boston, MA: Harvard Business School Press.

4   Conclusions

4.1   Reflections on the example of the float glass process

The float glass process, developed by Sir Alastair Pilkington and his team, is according to Smith (2010) a classic example of a process innovation that significantly improved the efficiency of previous glass production processes. Figure 4.1 is a picture of the float glass production team with Celfyn Thomas sitting centre front.

In the process, glass is produced by drawing it across a molten bed of tin (Quinn, 1991). Before the introduction of the process, glass was of poor quality and expensive due to the method of producing a flat glass surface through grinding and polishing, and was used for shop and office windows (Smith, 2010). Through the float glass process the need to grind and polish, which was time-consuming, was removed, and resulted in a considerable reduction in cost (Smith, 2010). This led to property developers and architects, who in the past were restricted by cost, now being able to use large sheets of glass for the construction of new buildings (Smith, 2010). As a result, the construction of buildings during the last 50 years has seen airports, hotels, office blocks and shopping malls using large glass areas (Smith, 2010). Tushman and Anderson (1986), with regard to managing strategic innovation and change (Tushman and Anderson, 2004), have cited Sir Alastair Pilkington and his team, with the development of the float glass process for manufacturing glass an example of a discontinuity where technological change is a "bit-by-bit cumulative process until it is punctuated by a major advance" (Tushman and Anderson, 1986, p. 41). As a major advance, float glass was highly disruptive but was followed by a period of relative stability (Smith, 2010). Further innovations followed this major advance, which were incremental and resulted in further product improvements instead of changes that were significant (Smith, 2010).

Figure 4.1 Picture of the float glass production team
Source: James Edward Celfyn Thomas' archive.

According to Abernathy and Utterback (1978), the need to improve manufacturing processes acts as a stimulus for innovation and this occurred in the
glass industry, which at a point of maturity in its evolution and with pressures
for a reduction in costs acted as a stimulus to undertake process innovation
to develop an efficient process. This is exemplified by the development of
the float glass process in the 1950s by Sir Alastair Pilkington and his team,
where these conditions existed (Smith, 2010). The manufacture of plate glass
at the time, especially for the construction industry, was time-consuming and
expensive, requiring glass sheets to be ground and polished to obtain a flat
surface; the float glass process removed these stages of production (Smith,
2010). Glass was drawn from a furnace over a molten bed of tin, and although
completing the innovation took a number of years, it resulted in faster glass
production at lower cost (Smith, 2010). Due to this major advancement in
the production of glass, Pilkington licensed the process to other major glass
manufacturers, resulting in float glass still being used on a wide basis (Smith,
2010). Float glass, as an innovation, arose from a process need through a
production restriction, which was an innovation source (Smith, 2010). It was
this restriction that provided an innovation stimulus (Smith, 2010).

Twiss (1979) makes five observations on the float glass process: (1) radical innovation transformed the industry and competitive position of the firm, having a major impact on competitors who were at first unwilling to license the process; (2) successful innovation offered substantial economic benefits through manufacturing cost reduction; (3) the new process was realised through the work of the project champion in a company which supported him; (4) difficulties, risks and uncertainties were high and financial implications at the outset could not be determined; and (5) theoretical knowledge was advanced through technical problems being overcome. An environment was present where entrepreneurial drive and individual creativity flourished involving a project champion, innovative environment, creativity, project management, evaluation of techniques, relationship to corporate objectives, market orientation and good decision-making regarding the new process (Twiss, 1979).

4.2   Licensing and the early float glass start-ups

Three important aspects to licensing are knowledge, access to finance and motivation (Smith, 2010), and this was the case with the float glass process. In fact, in order for Pilkington to develop its revolutionary float glass process for manufacturing glass, it based market leadership on licensing (Smith, 2010). Through following this licensing strategy, the company was provided with an income and it prevented alternative processes being developed by other companies. It was highly successful, with the first foreign license issued in 1962 to the Pittsburgh Plate Glass Company (Henry and Walker, 1991) and the first float glass start-up. Table 4.1 provides an overall list of the first licenses (Grundy, 1990) leading to the float glass start-ups in the early 1960s.

Table 4.1 First licenses leading to the float glass start-ups in the early 1960s

Date	License
July 27, 1962	Pittsburgh Plate Glass
December 14, 1962	Glaverbell S.A.
December 14, 1962	Glasses de Boussois
April 3, 1963	Fabrica Pisna de Specchi Lastre Colate di Verro Della
April 10, 1963	Glaceries de St Roche
April 19, 1963	Libbey Owens Ford
January 6, 1964	Vereinigte Glassworks GmbH
January 6, 1964	Compagnie de St Gobain
March 6, 1965	Asahi Glass Co. Ltd.
March 6, 1964	Nippon
June 3, 1964	Cristaleria Espanola
July 29, 1964	Ford Motor Company
March 29, 1965	Vidrio Plano de Mexico

Source: Grundy (1990).

It had taken seven years of R&D and millions of pounds to develop the float glass process for the manufacture of glass in the 1950s by Pilkington, based in Lancashire (Smith, 2010). This involved building small-scale pilot plants and a full-scale plant for the process, and it took a year for commercial-quality glass to be produced (Smith, 2010). The fact that flat glass was made by floating it on a bed of tin meant that rather than being based on discovery through curiosity-oriented research, it was a technological discovery (Jevons, 1973). Tracing the cause of the difficulties that arose was helped by science, and diagnosis of surface bloom on the glass through dissolved tin depended on techniques and concepts from basic research, a number of which could have been curiosity oriented (Jevons, 1973).

It has been argued that offering a manufacturing license for an innovation can disarm the innovator by diverting their research effort, and this approach was followed by Pilkington Brothers with the float glass process to realise considerable license fees (Twiss, 1979). A number of observers have argued that rather than follow this licensing strategy, Pilkington could have taken an offensive marketing strategy for their own products to achieve a larger market (Twiss, 1979).

4.3   Leadership

Aspects of leadership involved in the float glass process not only included individual leadership but also company leadership. This is shown by Pilkington, who were technology leaders with other firms being followers (Tidd et al., 1998). According to Teece (1986), the ability of a company to gain the benefits of investment in a technology is dependent on two aspects. First, there is the firm's ability to transform a technological advantage into processes and products that are commercially viable, and second, there is the ability of the firm to keep its advantage by fending off imitators (Tidd et al., 1998). In order to achieve this Pilkington defended its technological leadership through effective patent protection and became a competitive winner (Tidd et al., 1998).

On the individual leadership level within the company, the existence of key enabling figures is important, and through these people it is possible to overcome the complexity and uncertainty with an innovation in order to prevent a promising invention or innovation faltering before it can be released to the world outside the company (Tidd et al., 1998). In order to overcome such a problem it is helpful to have a key individual, and/or a team of people who are able to champion and enable an innovation project through the hurdles presented within an organisation (Nayak and Ketteringham, 1986; Wheelwright and Clark, 1992; Henry and Walker, 1991; Kidder, 1981). The project champion for the float glass process was Sir Alastair Pilkington, supported by key figures.

It has been noted by Fielden (Thring and Laithwaite, 1978) that the float glass process would never have been successful if the innovator had not been a Pilkington family member, because the process did run into many problems, and pessimists would have terminated the development before solving critical problems. According to Trott (2002), numerous company leaders have taken responsibility for technological innovation at a personal level in their company, and this was the case with Alastair Pilkington in the 1950s, who continued to push the float glass process against considerable resistance and eventually became successful.

4.4 Diffusion

The diffusion of the float glass process follows a pattern similar to the centre-periphery model (Schon, 1971), which rests on three basic assumptions:

1 The technology to be diffused exists prior to its diffusion,
2 Technology diffusion takes place from the source outwards to other businesses; and
3 The support of technology diffusion involves incentives, provision of resources and training.

This is shown in Figure 4.2 in terms of the early float glass start-ups.

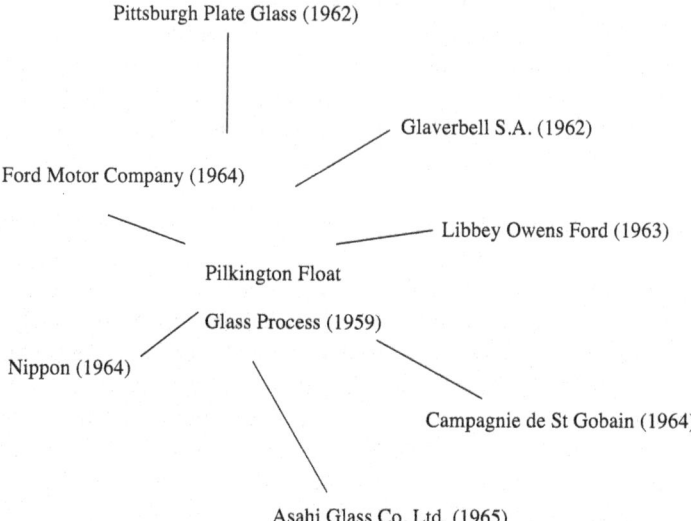

Figure 4.2 Early float glass start-ups

Diffusion took place from the source of the technology (Pilkington) through channels by a "diffuser" (float glass start-up teams), using a transfer mechanism (process start-up), to the business (licensee). The effectiveness of the start-up process depended upon the resources available to the licensee to enable the start-up, the efficiency of the start-up team and the process start-up involved, and the ability of the licensee to acquire the technology.

4.5   The float glass process and the importance of successful early start-ups to new process diffusion

According to Langrish et al. (1972), the float glass process was based on a technological discovery since it could not be considered to be the application of a discovery made by curiosity-oriented scientists or academics. The use of molten tin as a concept for the manufacture of flat glass had been patented in the United States in 1902, and it took £4 million of development work and seven years for the announcement in 1959 of the float glass process (Langrish et al., 1972). In recognition of the development of the process Pilkington Brothers won the Queen's Award for Industry. Following the process being developed empirically to a stage for the production of acceptable glass, in order to increase process understanding scientists were deployed to resolve problems of a technological nature, an example being tracing a layer of tin dissolved in the glass as the cause of a surface bloom (Langrish et al., 1972). This work used certain analytical techniques and basic concepts, possibly from curiosity-oriented research, but the science technology connections happened after initial process development, and the float glass process is an example of the use of science to solve an industrial problem (Langrish et al., 1972). It is interesting to note what David Pilkington remarked in a letter to Alun Thomas in March 2005, remembering the work of Alastair Pilkington and the contribution made by Celfyn Thomas:

> What a wonderful time it was when Alastair invented the "Atomic Bomb" of the glass industry.
>
> It was one thing to have the idea, it was quite another to make it work. But, to make it work on a pilot plant was quite another thing to turning it into a commercial production. This was where your father came in. What a man!
>
> What grit and courage and danger. People were injured – usually burned in the effort to make it work. Everything was red-hot. There hot gases emerging from the float bath, which had to be always under pressure. Molten tin was a hazard particularly when it heated, as it did.
>
> Alastair would visit at least twice a day and decide what to do next, but your father had to stay there all the time sometimes day and night and, of course weekends, making it work.
>
> (Pilkington, 2005)

Following these developments, the early float glass start-ups managed by Celfyn Thomas were important for the diffusion of the process. This shows that it was not only important for the invention of the process, development through its innovation, but for it to be made to work and for it to be successfully diffused, as was the case with the early float glass start-ups.

References

Abernathy, W. J. and Utterback, J. (1978) Patterns of industrial innovation, in M. L. Tushman and W. L. Moore (eds.), *Readings in the Management of Innovation*, New York: HarperCollins, pp. 97–108.

Grundy, T. (1990) *The Global Miracle of Float Glass: A Tribute to St Helens and Its Glass Workers*, St Helens: Chalon Press.

Henry, J. and Walker, D. (1991) *Managing Innovation*, London: Sage.

Jevons, F. R. (1973) *Science Observed: Science as a Social and Intellectual Activity*, London: George Allen & Unwin.

Kidder, T. (1981) *The Soul of a New Machine*, Harmondsworth: Penguin.

Langrish, J., Gibbons, M., Evans, W. G. and Jevons, F. R. (1972) *Wealth from Knowledge: A Study of Innovation in Industry*, London: Macmillan.

Nayak, P. and Ketteringham, J. (1986) *Breakthroughs: How Leadership and Drive Create Commercial Innovations That Sweep the World*, London: Mercury.

Pilkington, D. (2005) *Letter to Alun Thomas* (see Appendix 1).

Quinn, J. B. (1991) *The Strategy Process: Concepts, Contexts, Cases*, Englewood Cliffs, NJ: Prentice Hall.

Schon, D. A. (1971) Chapter 4, Diffusion of innovation, in *Beyond the Stable State*, London: Temple Smith, pp. 80–115.

Smith, D. (2010) *Exploring Innovation*, 2nd edn, London: McGraw-Hill.

Teece, D. (1986) Profiting from technological innovation: Implications for the integration, collaboration, licensing and public policy, *Research Policy*, 15, 285–305.

Thring, M. W. and Laithwaite, E. R. (1978) *How to Invent*, London: Macmillan.

Tidd, J., Bessant, J. and Pavitt, K. (1998) *Integrating Technological, Market and Organizational Change*, Chichester: John Wiley & Sons.

Trott, P. (2002) *Innovation Management and New Product Development*, 2nd edn, Harlow: Prentice Hall.

Tushman, M. L. and Anderson, P. (1986) Technology discontinuities and organisational environments, *Administrative Science Quarterly*, 31, 439–465.

Tushman, M. L. and Anderson, P. (2004) *Managing Strategic Innovation and Change*, 2nd edn, New York: Oxford University Press.

Twiss, B. C. (1979) *Managing Technological Innovation*, London: Longman.

Wheelwright, S. and Clark, K. (1992) *Revolutionising Product Development*, New York: Free Press.

Appendices

Appendix 1

Letter from David Pilkington
to Alun Thomas

I did try to telephone you, but Wolfgave seemed to have got the wrong number. 118888 said you were ex directory.

Tel: (01737) 764344
Fax: (01737) 780193

6 Cherry Green Close,
Oaklands Park,
Redhill, Surrey,
RH1 6RY

17/03/05

Dear Alan,

I was so sorry to hear about your father. It's like the end of an era, there are so few of us left who remember the glory days. I am 80 this year.

What a wonderful time it was when Alastair invented the "Alan Boats" of the glass industry.

It was one thing to have the idea, it was quite another to make it work. But, to make it work on a pilot plant was quite another thing to turning it into a commercial production. This was where your father came in. What a man!

What grit and courage and changes. People were injured — usually burned in the effort to make it work. Everything was red-hot. There were hot gases escaping from the float bath, which had to be always under pressure. Molten tin was a hazard particularly when it leaked, as it did.

Alastair would visit at least twice a day and decide what to do next, but ~~your father~~ had to stay there all the time sometimes day and night and, of course, weekends, making it work.

The other "big occasion" when your father came into his own was the 1970 strike. All production stopped. Chaos. Nobody even the strikers knew what was going on or how to solve it. Again your father showed courage, individual initiative, wisdom, originality. And humour.

This was more difficult than the development of float, because the rewards were not so obvious.

I remember my father dying and what a huge loss that was, so I send you my comfort and consolation, as much as I can.

Life moves on inexorably. We have to accept it, especially when the person concerned has had a long fruitful life.

My best wishes and prayers are with you and the family and particularly your mother.
          Yours ever
                    David Pilkington

Appendix 2

Letter from Buckingham Palace following royal visit

COPY

D 107 8/6

Cabinet (6¼" x 4¾")

3/- each

BUCKINGHAM PALACE

29th May, 1961.

Dear Sir Harry,

The Queen and The Duke of Edinburgh so much enjoyed their visit to your Cowley Hill Works on Thursday last, and have commanded me to thank you and the members of your Board for the excellent arrangements which you had made for them to see the Plate Glass and Float Glass processes, and to meet several of the staff of the firm engaged in operating them. It was a most interesting experience for them to see the new Float Glass process, which is such a striking development in the manufacture of Plate Glass.

Her Majesty and His Royal Highness are specially grateful to you for having devised such a thorough and unhurried tour for them; and Lady Leicester, Captain Harvey and I join in expressing our thanks to you for having arranged for everything to be explained also to us so well by other members of the firm. I have been to several factories with The Queen and can say without hesitation that I cannot remember a better arranged visit than this one.

Yours sincerely,

Edward Ford.

Sir Harry Pilkington.

C/ All Departmental Managers, C.H.W.

Source: James Edward Celfyn Thomas' archive

Appendix 3

Example of notes made from Jones (1956)

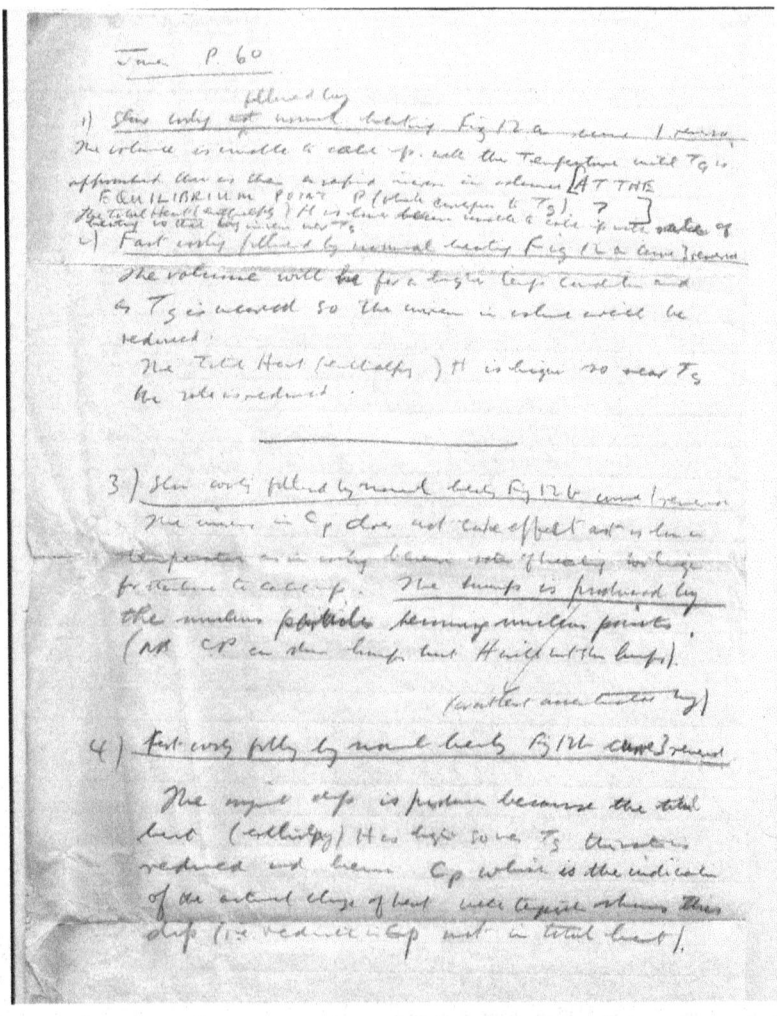

Source: James Edward Celfyn Thomas' archive

Index